The Laboratory
ZEBRAFISH

The Laboratory Animal Pocket Reference Series

Series Editor
Mark A. Suckow, D.V.M.
Freimann Life Science Center
University of Notre Dame
Notre Dame, Indiana

Published Titles

The Laboratory Canine
The Laboratory Cat
The Laboratory Guinea Pig
The Laboratory Hamster and Gerbil
The Laboratory Mouse
The Laboratory Nonhuman Primate
The Laboratory Rabbit, Second Edition
The Laboratory Rat
The Laboratory Small Ruminant
The Laboratory Swine, Second Edition
The Laboratory Xenopus sp.
The Laboratory Zebrafish

A Volume in The Laboratory Animal Pocket Reference Series

The Laboratory
ZEBRAFISH

Claudia Harper
Preclinical Amgen, Inc.
Cambridge, Massachusetts, U.S.A.

Christian Lawrence
Children's Hospital
Boston Massachusetts, U.S.A.

CRC Press
Taylor & Francis Group
Boca Raton London New York

CRC Press is an imprint of the
Taylor & Francis Group, an **informa** business

CRC Press
Taylor & Francis Group
6000 Broken Sound Parkway NW, Suite 300
Boca Raton, FL 33487-2742

© 2011 by Taylor and Francis Group, LLC
CRC Press is an imprint of Taylor & Francis Group, an Informa business

No claim to original U.S. Government works

International Standard Book Number: 978-1-4398-0743-9 (Paperback)

Library of Congress Cataloging-in-Publication Data

Harper, Claudia.
    The laboratory zebrafish / Claudia Harper and Christian Lawrence.
        p. ; cm. -- (Laboratory animal pocket reference series)
    Includes bibliographical references and index.
    Summary: "Second only to the mouse and rat, Zebrafish are the most popular animal model in biomedical research. Rapid embryonic development and transparent organogenesis, give Zebrafish unique advantages in the study of biological pathways, vertebrate development, carcinogenicity, drug development, genomics, gene function, mutagenesis screening, and toxicology. Providing a quick reference source for basic information and common procedures, this guide is designed for a range of experience levels. It covers all aspects pertaining to the use of these organisms including humane care and management, husbandry, compliance, technical procedures, veterinary care, housing, and water quality management"--Provided by publisher.
        ISBN 978-1-4398-0743-9 (pbk. : alk. paper)
    1. Zebra danio--Handling--Handbooks, manuals, etc. 2. Fish as laboratory animals--Handbooks, manuals, etc. 3. Fish culture--Handbooks, manuals, etc. 4. Zebra danio--Genetics--Handbooks, manuals, etc. I. Lawrence, Christian, 1972- II. Title. III. Series: Laboratory animal pocket reference series.
    [DNLM: 1. Animals, Laboratory--Handbooks. 2. Zebrafish--Handbooks. QY 60.F4]

QL638.C94H37 2011
639.3'7482--dc22   2010026453

Visit the Taylor & Francis Web site at
http://www.taylorandfrancis.com

and the CRC Press Web site at
http://www.crcpress.com

To our families

Evan and Skylar Mead

Kim, Drew, Julia, and Skippy Lawrence

Contents

preface

The rise of the zebrafish as an experimental animal has been a truly dramatic one. In the relatively brief span of only a few decades, it has gone from being mainly a hobby fish to a mainstream model animal employed by scientists to study everything from stem cells to the basis of behavioral changes induced by drug addiction. This rapid advance has been fueled largely by numerous and impressive advances in technology, along with detailed characterization of the animal on a genetic and molecular level. These developments have allowed scientists to leverage the many advantages of the zebrafish system to address many important questions in biology and human genetics and disease.

Given all of this, it may come as a surprise to many individuals that there are few accepted and established standards for husbandry, management, and care for the fish in laboratory settings. Practices are variable and rudimentary across the community, and while a number of references concerning the management, care, and husbandry of zebrafish do exist, they are disparate in nature, and are not comprehensive in terms of subject content. This information gap represents a major challenge for the growing number of people charged with the care, regulatory oversight, biological model standardization, and health management of zebrafish in research facilities across the globe.

To this end, the goal of this handbook is to provide managers, veterinarians, investigators, technicians, and regulatory personnel with a concise but comprehensive reference on zebrafish biology, care, husbandry, and management. The aim of the book is not to set standards, but rather to arm those working with the fish with scientifically grounded principles and fundamental information that can be used to design sound fish care programs.

This handbook is organized into seven chapters: "Biology" (Chapter 1), "Husbandry" (Chapter 2), "Life Support" (Chapter 3), "Management" (Chapter 4), "Veterinary Care" (Chapter 5), "Experimental Methodology" (Chapter 6), and "Resources" (Chapter 7). The final chapter, "Resources," provides the zebrafish user with lists of sources of additional information on the zebrafish model, as well as key references, professional organizations, and suppliers of equipment and supplies used in zebrafish husbandry and care. These lists are not exhaustive, and do not imply endorsement of referenced vendors or organizations; rather, they are simply good starting points for those working with the fish, especially those who are new to the model.

authors

Claudia Harper, DVM, Dipl. ACLAM, is currently a director at Amgen. Dr. Harper was previously a senior clinical veterinarian at Massachusetts General Hospital Center for Comparative Medicine, and a senior scientist at PharmaMar USA. She completed her veterinary degree at Tufts University and was a post-doctoral fellow in the Division of Comparative Medicine at the Massachusetts Institute of Technology.

Her experience includes research, and clinical and diagnostic work with aquatic biomedical research models in academia and the pharmaceutical industry. She played a key role at MIT in identifying, naming, and describing the aquatic *Helicobacter* species *H. cetorum*. She has presented original research and given lectures nationally and internationally. She has authored and coauthored numerous scientific articles.

Dr. Harper's clinical experience in aquatic animal health includes working with biomedical research models and aquaculture species. She was the fish health columnist for *Aquaculture Magazine* from 2002 until 2007 and is the co-founder and a past president of the Zebrafish Husbandry Association (ZHA).

Christian Lawrence, M.S., manages the Aquatic Resources Program at Children's Hospital Boston, which is home to one of the largest and most active zebrafish research programs in the world.

He earned his master of science in biology from the University of Massachusetts–Boston in 2006, where his graduate research focused on the environmental and genetic controls of sexual differentiation in the zebrafish. He also has a bachelor of science in wildlife conservation biology from Arizona State University and a bachelor of arts in communications from Seton Hall University.

Mr. Lawrence has worked with and managed zebrafish for nearly 10 years, spending time at Harvard University, the Marine Biological Laboratory, and Brigham and Women's Hospital before moving on to Children's Hospital Boston in 2008. He has authored a number of scientific papers on the biology, husbandry, and management of zebrafish. He is past president of the Zebrafish Husbandry Association (ZHA) and currently serves as the coordinator of the ZHA working groups program.

acknowledgments

We would like to express our gratitude to our colleagues who have supported us and contributed in the creation of this book. Thank you to Dr. Mike Kent and Dr. Trace Peterson from Oregon State University and Dr. Mike Ballinger and Eric Matthews from Amgen. We also thank the Aquatic Resources Program team at Children's Hospital Boston, as well as Yi Zhou, Austin Bailey, and others for comments and suggestions on sections of the manuscript.

biology

introduction

The zebrafish, *Danio rerio*, is a small tropical freshwater fish that has long been a favorite of the fish hobbyist. Many of the same characteristics that have popularized this species in the pet trade, including its tolerance of a broad range of environmental conditions and the ease with which large numbers of the animals can be bred and reared in captivity, have also made it attractive to scientists interested in using the fish to study various aspects of biology. Indeed, the zebrafish has been a mainstay in the laboratory since the mid-1900s (Laale 1977).

In the 1980s, a group led by Dr. George Streisinger at the University of Oregon performed revolutionary, pioneering work that would only a short time later allow scientists to employ techniques that had previously been feasible only in invertebrates to study the genetics of development in higher organisms. Using methods developed by Streisinger and his colleagues as the basis for their approach, groups in Boston and Germany performed two large-scale genetic screens that produced thousands of developmental mutants that proved to be invaluable tools for understanding the molecular basis of development. These landmark studies, first published in the early 1990s, formally ushered zebrafish into the mainstream as a classic developmental model system. Over time, the use of the fish was extended to other fields of science, including human health and disease, toxicology, behavior, and evolution.

Today, less than two decades after those first screens were published, the zebrafish has become a pre-eminent laboratory animal, on par with the mouse and rat in terms of its overall value for science and its prevalence in research programs. The following sections in this first chapter of the book describe important biological features of the zebrafish, as an understanding of the organism as a whole functioning animal is of paramount importance to its use as a model in any scientific setting.

natural history

Zebrafish, native to South Asia, are distributed primarily throughout the lower reaches of many of the major river drainages of India, Bangladesh, and Nepal (Spence et al. 2008) **(Figure 1)**. This geographic region is characterized by its monsoonal climate, with pronounced rainy and dry seasons. Such seasonality in rainfall exerts profound and sometimes drastic effects on environmental conditions in zebrafish habitats, both in terms of the physical and chemical qualities of the water and in the abundance of resources.

Zebrafish are primarily a floodplain species, and are most typically encountered in shallow, standing, or slow-moving bodies of water with submerged aquatic vegetation and a silt-covered substratum (Spence et al. 2008) **(Figure 2)**. The water quality in these habitats can vary widely. For example, the ranges of pH, conductivity, and temperature recorded at a number of sites where zebrafish have been collected in the wild are 5.9–8.5, 10–2000 µS, and 16–38°C, respectively (Engeszer et al. 2007; Spence et al. 2006). These differences, which can be explained by both seasonality and geography, suggest that zebrafish are well adapted to such fluctuations and likely account for their tolerance of a broad range of conditions in captivity.

In the wild, zebrafish feed mainly on a wide variety of zooplankton and insects (both aquatic and terrestrial), and to a lesser extent algae, detritus, and various other organic materials (McClure et al. 2006; Spence et al. 2007). As they occupy the whole of the water column, they feed largely within this zone, but will readily take items on the surface and the bottom (Spence et al. 2008). This supposition is supported by results of gut content analyses reported from wild-collected samples (Spence et al. 2007), as well as by observations made of fish feeding in laboratory settings.

Zebrafish are a shoaling species, most often occurring in small schools of 5–20 individuals (Pritchard et al. 2001), although groups of much larger numbers have also been reported (Engeszer et al. 2007). Reproduction takes place primarily during the rainy months,

Fig. 1 Natural distribution of zebrafish. Black dots indicate occurrences, and major river systems are indicated. Reprinted with permission from Spence, R. et al. 2008. *Biological Reviews*, 83:13–34.

a period of resource abundance (Talwar and Jingran 1991). Fish spawn in groups during the early morning along the margins of flooded water bodies, often in shallow, still, and heavily vegetated areas (Engeszer et al. 2007). Females scatter clutches of eggs over the substratum, and there is no parental care. The eggs, which are demersal and non-adhesive, develop and hatch in 48–72 hours at 28.5°C. After hatching, larvae adhere to available submerged surfaces by means of specialized cells on the head (Laale 1977). In 24–48 hours post-hatch, they inflate their swim bladders and begin actively feeding on small zooplankton. Larval fish remain in these nursery areas as they develop, and move into deeper, open water as they mature and as floodwaters recede (Engeszer et al. 2007).

Fig. 2 Representative zebrafish habitat in Bangladesh. Photo courtesy of Carl Smith.

genetics

Zebrafish are diploid animals, with a haploid complement of 25 mostly metacentric chromosomes (Postlethwait 2006). The genome, which is being sequenced by the Welcome Trust Sanger Institute, is ~1.4 x 10^9 base pairs in length, a little less than half the size of the human genome. As of this writing in the spring of 2009, the sequence was nearly finished, with a scheduled completion target of late 2009 (Welcome Trust Sanger Institute 2009). The genome, which contains a predicted ~22,000 genes, is rich in repetitive sequences, with an estimated overall repeat content of 40% (Reichwald et al. 2009). This is one of many features that are shared between the zebrafish and human genomes.

There is, in fact, a relatively high degree of genetic similarity between zebrafish and humans, despite a long evolutionary separation of several hundred million years (Postlethwait 2004). Nearly 60% of the genes expressed in zebrafish can be annotated with orthologues in the human genome based on their protein sequence homology (Children's Hospital Boston Zebrafish Genome Project 2009). In many cases, there are two orthologous copies of human genes present in the genome of zebrafish (Postlethwait 2004). This is consistent with a known whole genome duplication event in teleosts that occurred 150–300 million years ago, after the divergence of the teleost and tetrapod lineages (Jaillon et al. 2004). Interestingly, it is estimated that approximately 30% of the gene duplications in ancient teleosts have been retained in modern zebrafish (Farber et al. 2003). These duplicated genes may share the function of the single ancestral gene (Kleinjan et al. 2008), but in some cases one or both have gained a new role (Schlueter et al. 2006).

Genetic and molecular similarity to humans is an important attribute of the zebrafish as a model organism for biomedical research and studies of vertebrate development (Lieschke and Currie 2007). Because zebrafish and human genes show a relatively high degree of sequence homology and synteny (Barbazuk et al. 2000), screens can be performed in fish to help define gene functions involved in human development, disease, and behavior (Patton and Zon 2001). Several genetic and physical maps, including meiotic and radiation-hybrid maps, have been constructed to assist zebrafish genetic studies. An integrated map of these maps, known as ZMAP, has also been assembled and contains a total of ~35,000 markers. These genetic tools are necessary for positional cloning, genome sequence assembly, and comparative genomic studies between species (Day et al. 2009).

genetic strains

Genetic strains of zebrafish used in research can be placed into one of three general categories: **wild-type, mutant,** and **transgenic**. **Wild-type** strains, sometimes referred to as "standard lines," are technically defined as closed breeding populations of fish that harbor no defined phenotypic mutations (Trevarrow and Robison 2004). Wild-type zebrafish are representative of "typical" fish in terms of their basic anatomy, physiology, and behavior **(Figure 3)**.

The various wild-type zebrafish strains in existence are often utilized for different experimental purposes. For example, strains such as AB and Tübingen (TU), both of which underwent selection to remove background mutations early in their history, are now the most widely used strains in zebrafish research, and are employed in most types of studies, including sequencing, genetic screens, gene expression studies, and transgenesis. The WIK and SJD strains are used almost exclusively for genetic mapping, while highly genetically diverse strains such as Ekkwill (EKW) are employed in experimental instances where genetic uniformity is not required. A list of some of

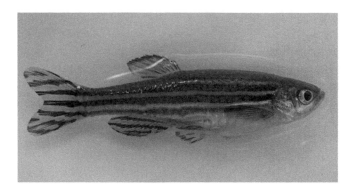

Fig. 3 Adult male wild-type zebrafish, AB strain.

the most prominent examples of zebrafish wild-type strains is provided in **Table 1.**

Most zebrafish wild-type strains are actually maintained as numerous small sub-populations with little or no programmatic interbreeding between them (Lawrence 2007). As a result, and because small populations are especially vulnerable to mutation, unconscious selection, natural selection, and especially genetic drift (Stohler et al. 2004), there is likely to be some degree of genetic divergence between the many subpopulations of any given zebrafish wild-type strain (Lawrence 2007).

A **mutant** zebrafish strain is one that carries one or more mutations **(Figure 4)**. Mutations, which are permanent changes in DNA sequence, can be spontaneous (Ren et al. 2002), but are most often induced. Mutations can be randomly induced into the genome of zebrafish by employing any one of a number of different approaches: chemical (e.g., ethylnitrosourea) mutagenesis (Mullins et al. 1994), irradiation (Walker and Streisinger 1983), or insertion of retroviruses or transposons into the genome (Wang et al. 2007; Nagayoshi et al. 2008). Randomly induced mutations are subsequently isolated and recovered by mating progeny of mutagenized animals and scoring them for desired phenotypes.

Targeted mutagenesis, or "knockouts," of specific zebrafish genes is also possible. This may be achieved through the use of zinc-finger nucleases (ZFNs), which are bioengineered molecules that induce mutations in a target DNA sequence (Doyon et al. 2008). Mutations in specific zebrafish genes may also be generated by TILLING (Targeting Induced Local Lesions in Genomes), a process in which a library of mutations randomly induced by chemical or insertional mutagenesis is screened in a high throughput manner for disruptions in specific genes (Moens et al. 2008; Lin et al. 2005).

TABLE 1: COMMON ZEBRAFISH WILD-TYPE STRAINS

AB (AB)	Widely used strain derived from pet shop stock in Oregon; initially screened to reduce transmission of embryonic lethal mutations
AB/Tübingen (AB/TU)	Hybrid cross of two widely used strains; maintained as single strain in closed population
Aquatica Tropicals (ZDR)	Strain maintained by Florida aquaculture supplier[a]
Ekkwill (EKW)	Strain maintained by Florida aquaculture supplier[a]
Nadia (NA)	Strain derived from wild-caught Indian stock; population maintained at the University of Oregon, USA
SJD (SJD)	Clonal line; established by sequential early pressure parthenogenesis; highly polymorphic with respect to other strains; excellent for mapping
Tübingen (TU)	Widely used strain derived from pet shop stock in Germany; initially screened to reduce transmission of embryonic lethal mutations; used by the Sanger Institute for the Genome Sequencing Project
Tupfel Longfin (TL)	Strain homozygous for leo[t1] and longfin[dt2] mutation used as wild-type strain; useful phenotypic markers
WIK (WIK)	Polymorphic with respect to other strains; used extensively for mapping

[a] Fish typically reared and maintained in ponds, greenhouses, or indoor production facilities.

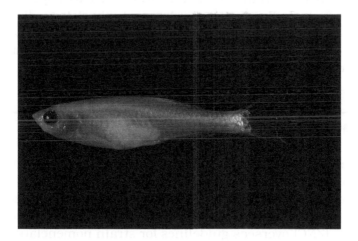

Fig. 4 Adult female *casper* mutant. The *casper* fish is homozygous for two mutations that result in extreme hypopigmentation. Photo courtesy of Richard White.

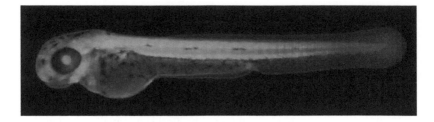

Fig. 5 Transgenic zebrafish larva. Larval fish carrying transgene that expresses green fluorescent protein in all cells of the body. Photo courtesy of Christian Mossiman.

Transgenic zebrafish possess DNA (i.e., a gene or a part of a gene) from another species that has been deliberately inserted into their genome. Once a transgene has been integrated into the genome of an animal, it becomes heritable, and will be expressed in any progeny that carry one or two copies of the gene. Zebrafish strains that express stable transgenes may be readily generated and are widely used in zebrafish research. Many transgenic strains utilize fluorescent reporter genes, such as green fluorescent protein (GFP), to label specific cell or tissue types **(Figure 5)**.

There are many thousands of mutant and transgenic zebrafish strains in existence that are used to study various aspects of vertebrate biology, including organ development, physiology, disease, and behavior. Mutant and transgenic strains are maintained by selective breeding of animals that carry one or two copies of the mutation(s) or transgene(s).

nomenclature

Nomenclature is a fundamental element of the zebrafish model system. The use of the appropriate nomenclature identifies a fish in terms of its specific genetic makeup and distinguishes it from the numerous other zebrafish strains in existence. The Zebrafish Nomenclature Committee (ZNC) oversees guidelines for strain nomenclature, which is constantly evolving along with new and emerging methods for the creation of mutant and transgenic animals. A complete description of the most current set of rules for naming zebrafish strains can be found on the Zebrafish Information Network (ZFIN) website (http://zfin.org/zf_info/nomen.html#4.1). An abbreviated version of these guidelines is presented below:

Genes: The full names of zebrafish genes, as well as their corresponding gene symbol, are italicized. The letters of gene symbols are unique to that zebrafish mutation or gene except in the case of established orthologues, in which case it should match that of the orthologue. Genes that are members of gene families are sequentially numbered. Example: name: *dystrophin*; symbol: *dmd.*

Wild-type strains: The names of wild-type strains are non-italicized with the first letter of the name in uppercase; abbreviations approved by the ZNC are all uppercase and non-italicized. Example: name: Tuebingen; symbol: TU.

Mutant strains: Zebrafish strains containing spontaneous or chemically induced mutations in identified genes are designated using the appropriate symbol for the gene that contains the mutation, followed by a superscript line designation, which is a one-to-three-letter code specific to the institution where the mutant was isolated. The line designation code is followed immediately by a number, which typically (but not always) denotes the order in which the allele was identified at that particular institution. Example: *trim33* $^{m62/m62}$

1. The first part of the name indicates the standard abbreviated name of the gene that carries the mutation; in this case *trim33* refers to the zebrafish *tripartite motif-containing 33* gene.

2. The second part of the name identifies the genotype of the line; in this case the line carries two copies of a mutant allele of the *trim 33* gene that was isolated in a laboratory at Massachusetts General Hospital, *m62.*

Mutant strains with unidentified genes: Mutant loci for which the gene has not yet been identified are given temporary gene names that are related in some way to the observed phenotype. When the gene is eventually identified, it is renamed in accordance with standard nomenclature described above. Zebrafish mutant names must be unique to the mutant. Example: *clo*$^{m39/m39}$

1. The first part of the name is the abbreviation of the name given to the mutant loci, in this case, *cloche.*

2. The second part of the name identifies the genotype of the strain; in this case the strain carries two copies of the *cloche* mutant allele that was isolated in a laboratory at Massachusetts General Hospital, *m39.*

Mutant strains generated by insertion: Mutant strains generated by insertion of retroviruses, transposons, zinc-finger nucleases, etc., are designated in the same manner as zebrafish mutant strains with spontaneous or chemically induced mutations, except that the line designation is followed by a two-letter designation indicating the nature of the insertion or "knockout." Example: $dtl^{hi447Tg/hi447Tg}$

1. The first part of the name is the abbreviation of the name of the mutated gene, in this case the zebrafish *dtl* gene.

2. The second part of the name identifies the genotype of the strain; in this case the strain copies two copies of a mutant allele generated in a laboratory at the Massachusetts Institute of Technology, *hi447*. The mutation was generated by transgenic insertion of a retrovirus, which is designated *Tg*.

Transgenic strains: Transgenic zebrafish lines are denoted using a standard abbreviation for the mode of insertion, the inserted foreign gene or genes, and the laboratory designation. Example: *Tg(gata1:dsRed)sd2*

1. The first part of the name (*Tg*) denotes the mode of insertion, in this case a transgenic construct inserted by microinjection into a fertilized embryo at the one-cell stage.

2. The second part of the name refers to the inserted transgene, in this case the promoter of the zebrafish *gata1* gene fused to the fluorescent reporter *dsRed*.

3. The third part of the name refers to the laboratory designation assigned to the strain; in this case *sd2* indicates that it was generated in a laboratory at the University of California–San Diego.

behavior

Zebrafish are a shoaling species, showing a tendency to form mixed-sex schools in nature and captivity (Engeszer et al. 2008). This behavior is innate (Kerr 1962; Engeszer et al. 2007) and heritable (Wright et al. 2003), and individuals choose shoals upon the basis of both visual (Spence and Smith 2007) and olfactory (Gerlach and Lysiak 2006) cues.

Despite the fact that they are social animals, zebrafish frequently exhibit agonistic behavior, especially when mating and during the establishment of dominance hierarchies, which occur within and between the sexes (Spence et al. 2008). These interactions may be mediated by direct, physical aggression, such as biting, chasing, and circling, or indirectly via the release of repressive pheromones into the water by dominant fish (Gerlach 2006). These types of behaviors may promote stress in laboratory stocks, and should therefore be managed carefully.

Zebrafish are markedly diurnal, showing the highest levels of activity throughout the day, particularly during the early morning (MacPhail et al. 2009). They do sleep, most frequently, although not exclusively, at night (Yokogawa et al. 2007). This clear circadian pattern of activity governs many physiological, biochemical, and behavioral processes in the animal, and is highly dependant upon the establishment of a regular photoperiod when the fish are maintained in artificial settings. Interruptions in the light cycle may be highly problematic for zebrafish kept in the laboratory, especially with respect to maintenance of the reproductive cycle (Selman et al. 1993).

anatomy and physiology

Integumentary System

Unlike reptiles and mammals, fish do not have a keratin layer over the epidermis. Zebrafish skin is covered by cycloid scales and provides a protective physical barrier that is important for both osmoregulation and pathogen defense. In the early developmental stages of the larval zebrafish, cutaneous exchange meets both ionoregulatory and respiratory demands (Rombough 2002). Fish skin may be damaged by handling, fighting, physical trauma, predation, environmental irritants, and pathogens. Any of these events may lead to opportunistic microbial infections. Zebrafish are easily recognized by their horizontal stripes, which are formed by a pigment pattern of three types of pigment cells: melanophores, xanthophores, and iridophores (Hirata et al. 2003).

Musculoskeletal System

Zebrafish are classified as a bony fish. Their axial skeleton includes the vertebral column and the unpaired fins (Bird and Mabee 2003).

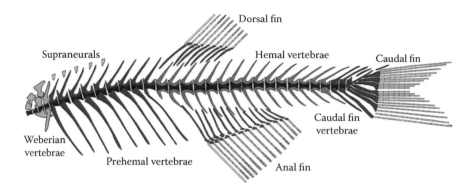

Fig. 6 Diagram of the zebrafish axial skeleton. Centra are black, the Weberian apparatus is green, supraneurals are light blue, precaudal vertebrae are red, caudal vertebrae are orange, the caudal fin skeleton is purple, and the dorsal and anal fin endoskeletons are blue. Bird, N. C., Mabee, P. M. 2003. 228, 3, 337–357. Reprinted with permission of John Wiley & Sons, Inc.

The vertebral column has two distinct sections: the precaudal and the caudal vertebrae. Zebrafish have three unpaired fins: the anal fin, the dorsal fin, and the caudal fin (Bird and Mabee 2003). They have two sets of paired fins: the pelvic fins and the pectoral fins. Zebrafish typically have 10 ribs (Bird and Mabee 2003).

Zebrafish have a Weberian apparatus, a trait that they share with other ostariophysan fishes. The Weberian apparatus is composed of a series of small bones located cranially to the precaudal vertebrae that transmit sound and vibrations from the swim bladder to the inner ear in zebrafish (Bird and Mabee 2003). A diagram of the zebrafish axial skeleton is shown in **Figure 6**.

Gastrointestinal System

The alimentary tract of the adult zebrafish includes the mouth, pharynx, esophagus, intestinal tract, and the anal opening, which is sometimes referred to as the proctoderm or urogenital pore (Ng et al. 2005; Wallace et al. 2005). Similarly to other fish belonging to the order Cypriniformes, zebrafish have a pair of teeth that are restricted to a single pair of pharyngeal bones (Stock 2007).

Zebrafish have taste buds that are chemosensory organs made of modified epithelial cells (Hansen et al. 2002). Taste bud cells assist the fish in deciding whether substances are edible. Taste bud cells are located on the lips, in the mouth and the oropharyngeal cavity, on the barbels and the head, and in some instances on the body surface (Hansen et al. 2002). Interestingly, the zebrafish taste buds first develop on the lips and gill arches between 4 and 5 days postfertilization, coincident with the time when the fish starts feeding (Hansen et al. 2002). The taste buds in the mouth, on the head, and on the barbels develop later (Hansen et al. 2002).

The intestine fills the abdominal cavity and has three functional divisions referred to as the anterior, mid, and posterior intestine (Ng et al. 2005; Wallace et al. 2005). The anterior portion, also known as the intestinal bulb, is larger and acts as the main location for lipid and protein digestion (Rombout et al. 1985). Other members of the Cyprinidae family have an intestinal bulb, but in them it is referred to as the pseudogaster (Harder 1975; Field et al. 2003). Fatty-acid-binding proteins are highly expressed in the anterior intestine of the zebrafish larva, indicating that fat absorption probably occurs in that intestinal region (Her et al. 2004; Mudumana et al. 2004; Wallace et al. 2005). It is also believed that the anterior intestine plays an important role is nutrient absorption, based on the high concentration of digestive enzymes in this region as well as the height of the epithelial folds (Her et al. 2004; Mudumana et al. 2004; Wallace et al. 2005).

There are several differences between the gastrointestinal tract of mammals and the zebrafish. The zebrafish does not have intestinal paneth cells, crypts of Lieberkuhn, a stomach, or a cecum (Ng et al. 2005; Wallace et al. 2005). The absence of a stomach indicates that acidification is not required for zebrafish digestion, and it is thought that the intestinal bulb functionally replaces the stomach (Wallace et al. 2005; Lieschke and Currie 2007).

The liver in fish plays an important role in energy storage and metabolism. Unlike the liver in mammals, the zebrafish liver does not have a lobular architecture. Rather, the liver is a bi-lobed organ divided into the right lobe and left lobe (Chu and Sadler 2009). The liver is positioned cranially and ventrally relative to the swim bladder (Chu and Sadler 2009). The zebrafish has an intrahepatic biliary tree and extrahepatic bile ducts and gallbladder (Matthews et al. 2004).

Respiratory System

The gills of teleost fish comprise four bilateral pairs of gill arches. Each gill arch gives rise to two parallel rows (hemibranches) of gill filaments, which are also called the primary lamellae. Rows of

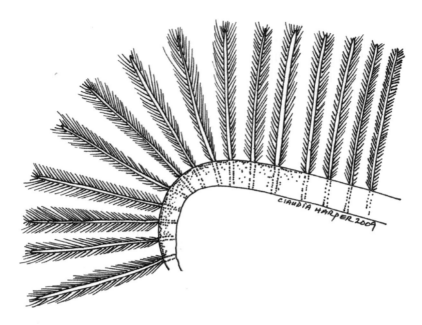

Fig. 7 Illustration of adult zebrafish gill anatomy. The arch-shaped gill arch supports the long primary lamellae from which the shorter secondary lamellae branch off.

secondary lamellae arise from the primary lamellae **(Figure 7)** (Jonz and Nurse 2008).

In the adult zebrafish, the gills are responsible for gas exchange, ion and water balance, excretion of nitrogenous wastes, and the maintenance of acid–base balance (Rombough 2002). The secondary lamellae are the gas-exchange structure of the gills in the adult fish. Mitochondria-rich cells, the ionocytes, are involved in ion exchange, acid–base balance, and ammonia excretion (Perry 1997; Wilson et al. 2000; Shih et al. 2008).

It is important to note that in zebrafish, the gills are not the primary site of gas and ion exchange until the fish has fully metamorphed into the juvenile stage (Rombough 2002). When the fish first hatches, both gas and ion exchange occur exclusively by diffusion across the skin. After a few days, ionoregulation shifts to the gills, but gas exchange still occurs across the skin. As the larva begins its transformation into the juvenile stage, the gills begin to assume some, but not all respiratory function, and so gas exchange happens at both sites. It only after metamorphosis has been completed that the gills become the primary site of both ionoregulation and respiration (Rombough 2002).

The gills are covered bilaterally with a protective bony plate called the operculum. Attached to the undersurface of the operculum is the pseudobranch, which is a modified gill arch with a single row of filaments. The function of this organ in zebrafish is undetermined. There are several hypotheses regarding the function of the pseudo-branch (Waser and Heisler 2005; Jonz and Nurse 2008; Mölich et al. 2009). It is believed that it may supply highly oxygenated blood to the optic choroid and retina of fish and may have thermoregulation and baroreceptor functions (Mölich et al. 2009; Waser and Heisler 2005). However, there are species-specific differences in regard to the role and function of the pseudobranch in fish (Jonz and Nurse 2008). For example, it seems that the pseudobranch does not play an important role in O_2 sensing in zebrafish and trout (Jonz and Nurse 2003, 2008, 2005).

Swim Bladder

Zebrafish have a swim bladder, a gas-filled organ that lines the dorsal portion of the abdominal cavity on the ventral side of the kidneys. Its location in the abdominal cavity is important since it impacts the fish's stability and buoyancy (Robertson et al. 2008) The swim bladder occupies 5.1+/−1.4% of total body volume in zebrafish, which significantly decreases the whole-body density of the animal (Robertson et al. 2008). The swim bladder has anterior and posterior chambers connected to each other by the ductus communicans **(Figure 8)**. The posterior chamber is connected to the esophagus by the pneumatic duct (Finney et al. 2006). Zebrafish can regulate the volume of gas that enters or leaves the swim bladder. This buoyancy-regulating ability is believed to be controlled by autonomic neurons (Finney et al. 2006).

In zebrafish, the anterior chamber functions primarily as an acoustic resonator to detect vibration, and the posterior chamber is used for buoyancy regulation as the fish moves through the water column (Finney et al. 2006; Robertson et al. 2007; Robertson et al. 2008). The zebrafish swim bladder is derived from the endoderm, and while it corresponds embryologically to the mammalian lung, the correspondence is not functional (Lieschke and Currie 2007).

Genitourinary System

Zebrafish kidneys are retroperitoneal and are divided into a cranial portion and a caudal portion, which are also known as the head

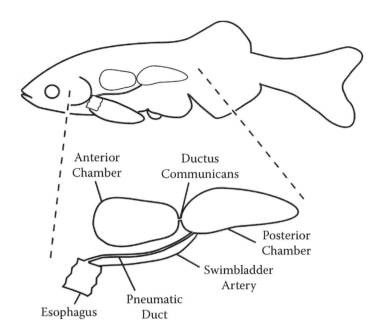

Anterior
Chamber

Ductus
Communicans

Posterior
Chamber

Swimbladder
Artery

Pneumatic
Duct

Esophagus

Fig. 8 Diagram of a lateral view of an adult zebrafish swim bladder showing its anatomical position. Finney, J. L. et al. 2006. 495, 587–606. Reprinted with permission of John Wiley & Sons, Inc.

and trunk kidney, respectively. In fish, cranial portion (head kidney) contains hematopoietic, lymphoid, and endocrine tissue, and the caudal portion (trunk kidney) contains nephrons and hematopoietic and lymphoid tissue (Reimschuessel 2001). Since fish do not have bone marrow, the anterior kidney is the main hematopoietic organ (Greenwell et al. 2003). Filtration occurs in both portions of the kidney. The glomerular anatomy and function of the zebrafish kidney is similar to that of humans. Unique features of the genitourinary system of the zebrafish include the absence of a bladder and a prostate gland (Lieschke and Currie 2007). Since zebrafish are small and do not have a bladder, urine collection is not easy.

Similarities between zebrafish and human reproduction include the molecular and embryological biology of the germ-cell development and the cellular anatomy of germ-cell organs (Lieschke and Currie 2007). The zebrafish male has a pair of testes, and the female has ovaries that contain oocytes (Deniz Koc et al. 2008). The fish fertilize the eggs externally, and the oocytes are surrounded by a chorion as opposed to a zona pellucida. The processes governing sex determination and gonadal differentiation in zebrafish are not well understood. Zebrafish do not appear to have sex chromosomes, and a master sex-determining gene has not been identified (Jørgensen et al. 2008). It is thought that sex determination is mediated by signals from a number of autosomal genes, and that the resultant process of gonadal differentiation can be influenced by environmental factors (Jørgensen et al. 2008; Orban et al. 2009). It is also believed that environmental factors may affect zebrafish reproduction and sex ratios (Orban et al. 2009). It has been shown that exposure to certain environmental endocrine disruptors can affect fecundity, sex differentiation, gamete development, and reproductive function of both male and female zebrafish (Xu et al. 2008).

Cardiovascular System

The zebrafish heart has two true chambers, called the atrium and ventricle. The anatomy of the heart also includes the sinus venosus and the bulbus arteriosus **(Figure 9)**. In zebrafish and other teleost fish, the bulbus arteriosus is a thick-walled chamber that connects the ventricle to the ventral aorta.

The venous return reaches the sinus venosus, which drains into the atrium surrounding the dorsal portion of the ventricle (Hu et al. 2001). The blood is then pumped from the ventricle through the bulboventricular orifice and into the bulbus arteriosus, which is connected to the ventral aorta (Hu et al. 2001). The bulbus arteriosus provides constant continuous blood flow to the gill arches, which are the primary site of blood oxygenation (Hu et al. 2001). The blood is then distributed through the dorsal aorta to the rest of the body (Hu et al. 2001; Moore et al. 2006).

The zebrafish embryo is particularly amenable for cardiovascular research since the heart is the most predominant organ in the developing embryo and it can easily be visualized by 72 hours post-fertilization. Also, the vascular tree is fully functional at that time (Milan et al. 2006; Briggs 2002). There are numerous similarities between the electrical properties of the human and the zebrafish heart, which make the zebrafish a suitable model for studying cardiac arrhythmias (Arnaout et al. 2007). Similarly to those in other vertebrate hearts, the cardiomyocytes from zebrafish embryos have voltage-gated sodium currents, L-type calcium currents, and T-type calcium and potassium currents (Baker et al. 1997; Brette et al. 2008; Nemtsas et al. 2009).

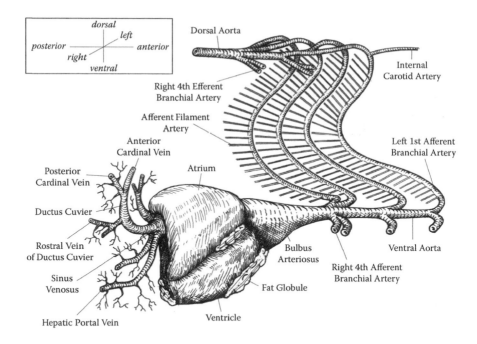

Fig. 9 An illustration of a posteroanterior view of an adult zebrafish heart and the major vasculature in the cardiac region. The atrium receives the venous return from the sinus venosus. The heart pumps the blood to the bulbus arteriosus along the definite chamber of atrium and ventricle. The ventricle forces the blood into the ventral aorta, which gives off paired vessels (afferent branchials) that arch upward between the successive gills to rejoin (efferent branchials) and form the dorsal aorta. The boxed area indicates the coordinates showing the orientation of the heart. Hu, N. et al. 2001. *The Anatomical Record*, 264, 1–12. Reprinted with permission of John Wiley & Sons, Inc.

Hematopoietic and Lymphoid System

Hematopoietic gene expression and function in zebrafish is conserved and similar to that in higher vertebrates (Brownlie et al. 1998; Hansen et al. 2002; Bertrand and Traver 2009; Kobayashi et al. 2009). A significant amount of work has been done on studying the hematopoietic system of the zebrafish (Barut and Zon 2000; Paw and Zon 2000; Onnebo et al. 2004; Amatruda and Patton 2008; Huang and Zon 2008; Bertrand and Traver 2009). Zebrafish have multiple hematopoietic cell types, including erythrocytes, myeloid cells, neutrophils, eosinophils, monocytes and macrophages, T cells, and

B-lymphocytes, which is similar to the case in humans. Hemostasis is also regulated by the coagulation cascade. Zebrafish have an innate immune system as well as an adaptive humoral and cellular immune system. Similarly to other fish, zebrafish do not have lymph nodes and they do not form germinal centers in the spleen (Lieschke and Trede 2009).

In zebrafish, the red blood cells and platelets are nucleated. The nucleated platelets in fish are referred to as thrombocytes. In embryonic zebrafish, primitive hematopoiesis occurs in the inter-mediate cell mass (ICM), which is a region ventral to the notocord and proximal to the axial vessel primordium (Leung et al. 2005). At 24 hours postfertilization, the ICM predominantly contains erythroid progenitor cells, whereas the yolk sac contains cells of myeloid lineage (Leung et al. 2005; Lieschke et al. 2002). Definitive hematopoiesis begins on day 4 (dpf) in the kidneys (Leung et al. 2005; Bennett et al. 2001). In mammals, the liver is an early hematopoietic organ, but this is not observed in zebrafish (Field et al. 2003).

Melanomacrophage centers (MMC), also known as pigmented macrophage aggregates, are a group of pigmented phagocytes that contain melanin and are found in the kidneys, spleen, and liver of fish, anurans, and reptiles (**Figure 10**) (Johnson et al. 2005; Deivasigamani 2007). Pigmented macrophage aggregates are present mostly when fish are stressed or diseased. In zebrafish, infected fish display MMC in the gills, liver, and kidneys (Sanders et al. 2003).

In vertebrates, the thymus is a critical lymphoid organ. In zebrafish, the thymic primordium is colonized by immature lymphoblasts a few days after fertilization (Willett et al. 1999).

Zebrafish produce both B and T cells, as well as the immuno-globulins IgM and IgZ (Danilova et al. 2005). A previously unknown immunoglobulin heavy-chain class was identified in zebrafish: immu-noglobulin zeta (Danilova et al. 2005). The immunoglobulin zeta gene (*ighz*) from zebrafish is expressed only in the kidney and thymus, the main lymphoid-producing organs in teleost fish (Danilova et al. 2005).

Endocrine System

There are great similarities between the endocrine system of zebrafish and that of mammals; however, one fundamental difference is that the zebrafish endocrine system responds to water parameter changes. For example, changes in salinity can upregulate or downregulate endocrine genes such as atrial natriuretic peptide, rennin, prolactin,

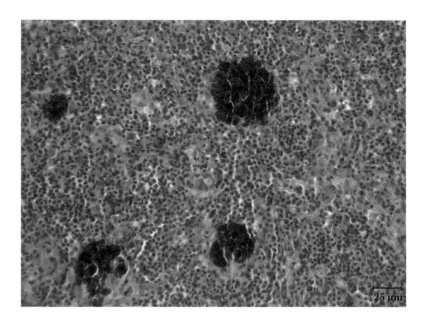

Fig. 10 Pigmented macrophage aggregates from zebrafish (Cypriniformes) spleen. Note individual melanomacrophages and non-encapsulated small aggregates of melanomacrophages around ellipsoids and scattered throughout the parenchyma. H&E 400X. Photo courtesy of Dr. Trace Peterson, Department of Microbiology, Oregon State University.

growth hormone, and parathyroid hormone (Hoshijima and Hirose 2007). Although this section does not focus on non-protocol-related changes associated with environmental parameters, the following summarizes some of the main anatomical and biological features of the endocrine system in zebrafish.

Pituitary gland

The pituitary expresses both prolactin and growth hormone in zebrafish (Hoshijima and Hirose 2007). Both proteins are primarily involved in osmoregulation in fish (Greenwell et al. 2003; Sakamoto and McCormick 2006; Hoshijima and Hirose 2007; Lieschke and Currie 2007). Prolactin controls the balance between salt and water by regulating the uptake of water and the retention of ions (Liu et al. 2006).

Adrenal gland

Like many other fishes, zebrafish do not possess a true adrenal gland. The functional counterpart of the mammalian adrenal gland in zebrafish is the interrenal gland, located in the anterior portion of the head kidney. The interrenal cells are mixed in with the chromaffin cells, as opposed to encapsulating them as seen in the mammalian adrenal gland (Liu et al. 2007). The interregnal cells secrete glucocorticoid and mineralocorticoid (McGonnell and Fowkes 2006). The chromaffin cells are functionally equivalent to the adrenal medulla (Liu et al. 2007).

Thyroid gland

Zebrafish do not have a compact thyroid gland. However, the main genes involved in the patterning and development of zebrafish thyroid tissue are conserved between zebrafish and mammals (Porazzi et al. 2009). The thyroid gland in the adult zebrafish is composed of loosely arranged follicles that are not encapsulated. Thyroid follicles are distributed around the ventral aorta, and found between the first gill and the heart (McGonnell and Fowkes 2006; Wendl et al. 2002). Zebrafish thyroid follicles produce thyroxine (T4) and tri-iodothyronine (T3). Zebrafish also produce thyroid hormone (TH) and have thyroid hormone receptors (TRs), both of which are required for zebrafish and mammalian development (Brown 1997; Liu et al. 2002).

Parathyroid glands

It is well established that the vast majority of fish do not have a parathyroid gland and do not produce parathyroid hormone (McGonnell et al. 2006). However, zebrafish have two parathyroid hormone genes and active peptides: the parathyroid hormone and calcium-sensing receptor gene. Both are expressed in the gills as opposed to in the parathyroid gland (Okabe and Graham 2004; McGonnell et al. 2006).

Pancreas

Zebrafish have an exocrine and endocrine pancreas that is scattered in the mesentery near the pylorus (Argenton et al. 1999; Yee et al. 2005; McGonnell and Fowkes 2006). The vertebrate pancreas has several roles: the exocrine gland is responsible for digestion and nutrition, and the endocrine gland is responsible for glucose homeostasis.

The morphology and ultrastructure of the exocrine pancreas (parenchyma), also known as the hepatopancreas, are highly conserved between zebrafish and mammals (Yee et al. 2005). In adult zebrafish, the exocrine pancreas is diffuse with ducts forming branches associated with acinar glands found among the intestinal loops; it is distributed near the small veins off the hepatic portal vein (Yee et al.

2005). The main pancreatic duct has branches that insert into the proximal intestine and transport the digestive enzymes secreted by the zymogen granule-containing acinar cells (Yee et al. 2005).

Similar to mammals, the endocrine pancreas of zebrafish has at least four cell types, which have functions similar to the mammalian pancreatic islet and are responsible for producing glucagon, insulin, and somatostatin (Kim et al. 2006). Interestingly, zebrafish and a number of other teleosts live in a hyperglycemic state (McGonnell and Fowkes 2006).

The endocrine pancreas (islet) contains the islets of Langerhans, which has insulin-secreting B-cells (Moss et al. 2009). The endocrine cells do not form an organ; instead they are scattered and organized as principal islets also known as Brockman bodies. In zebrafish, the Brockman bodies are located between the intestine and the liver, near the pancreatic duct, and consist almost entirely of endocrine cells (Milewski et al. 1998). Additional smaller islets can be found scattered along the intestine (Milewski et al. 1998).

Ultimobranchial gland

The ultimobranchial bodies in the adult zebrafish form two groups of follicles located at the transverse septum close to the sinus venosus of the heart and are responsible for producing calcitonin, which decreases serum calcium levels (Alt et al. 2006). In zebrafish, the ultimobranchial body is a separate organ from the thyroid gland (Alt et al. 2006).

Corpuscles of Stannius

In zebrafish, the corpuscles of Stannius, located dorsal to the distal tubule of the pronephric duct part of the renal system, are a unique cluster of cells that maintain calcium and phosphate homeostasis by producing stanniocalcin (previously called hypocalcin or teleocalcin) (Tseng et al. 2009; Wingert et al. 2007; Wingert et al. 2008). Stanniocalcin acts in conjunction with calcitonin, which decreases serum calcium levels. Corpuscles of Stannius have been identified in various fish, but it is unclear whether there is an analogous structure in mammals (Wingert et al. 2007; Wingert et al. 2008).

Urophysis

The urophysis is a unique neurosecretory organ located on the ventral aspect of the caudal end of the spinal cord. These bodies are composed of unmyelinated axons terminating on a capillary wall. The urophysis is composed of Dahlgren cells that synthesize two neuropeptides called urotensins I and II (Parmentier et al. 2006).

Pineal gland

The pineal gland in zebrafish is responsible for producing melatonin and for circadian regulation (Toyama et al. 2009). This light-sensitive structure is located in the anterior brain.

Nervous and Sensory Systems

Zebrafish eyes are very similar to mammalian eyes. The anatomy, histology, circuitry, and biochemistry of the eye are highly conserved among the different classes of vertebrates (Pujic and Malicki 2004; Soules and Link 2005; Fadool and Dowling 2008). In contrast to rodents, which are nocturnal animals, zebrafish are diurnal and their retinas have a large number of diverse cone subtypes as well as rods (Fadool and Dowling 2008).

Zebrafish have inner ears but no external ears (Whitfield et al. 2002). In vertebrates the auditory system relies on sound in the inner ear hair cells (HCs) being transduced into neural signals that are then transmitted to the brain (Tanimoto et al. 2009). The maculae, the cristae, and the organ of Corti all function to detect motion and sound (Tanimoto et al. 2009). In fish and frogs, the macular organs contain receptor hair cells and crystalline deposits of otoconia. The latter, more commonly referred to as otoliths, are important in detecting acceleration and gravity and are involved in hearing in lower vertebrates (Popper and Fay 1993; Tanimoto et al. 2009).

Similar to amphibians, the lateral line in fish is a sensory system that enables the animal to detect and respond to changes in water motion. The lateral line in zebrafish consists of sensory hair cells, known as mechanosensory neuromasts, that are embedded within the skin in lines running lengthwise on both sides of the body (McHenry et al. 2009). The lateral line is involved in a variety of behaviors, including predator evasion, schooling, and sexual courtship (Ghysen and Dambly-Chaudiere 2004; McHenry et al. 2009).

clinical chemistry and hematology

Little work has been done to establish the normative blood values for zebrafish. There are only a few articles published on this topic focusing

TABLE 2: ADULT ZEBRAFISH CLINICAL CHEMISTRY NORMATIVE VALUES

	Mean ± SD	Range
Albumin (g/dL)[1]	3.0 ± 0.2	2.7–3.3
Alkaline phosphatase (IU/L)[1]	2.0 ± 4.5	0.0–10.0
Alanine aminotransferase (IU/L)[1]	367.0 ± 25.3	343.0–410.0
Amylase (IU/L)[1]	2331.4 ± 520.6	1898.0–3195.0
Total bilirubin (mg/dL)[1]	0.38 ± 0.1	0.2–0.6
BUN (mg/dL)[1]	3.2 ± 0.4	3.0–4.0
Calcium (mg/dL)[1]	14.7 ± 2.3	12.3–18.6
Phosphorus (mg/dL)[1]	22.3 ± 1.5	20.3–24.3
Creatinine (mg/dL)[1]	0.7 ± 0.2	0.5–0.9
Glucose (g/dL)[1]	82.2 ± 12.0	62.0–91.0
Potassium (mEq/L)[1]	6.8 ± 1.0	5.2–7.7
Total protein (g/dL)[1]	5.2 ± 0.5	4.4–5.8
Globulins (g/dL)[1]	2.1 ± 0.6	1.3–2.8

[1] Murtha JM, Qi W, Keller ET. 2003. *Comp Med.* 53(1):37–41. With permission.

TABLE 3: ADULT ZEBRAFISH WHITE BLOOD CELL DIFFERENTIAL COUNT NORMATIVE VALUES

	Mean ± SD	Range
Lymphocytes[1]	82.95 ± 5.47	71.0 – 92.0
Monocytes[1]	9.68 ± 2.44	5.0 – 15.0
Neutrophils[1]	7.10 ± 4.75	2.0 – 18.0
Eosinophils[1]	0.15 ± 0.53	0.0 – 2.0
Basophils[1]	0.13 ± 0.40	0.0 – 2.6

[1] Murtha JM, Qi W, Keller ET. 2003. *Comp Med.* 53(1):37–41. With permission.

on zebrafish. Currently available data are compiled in **Table 2**, which addresses the adult zebrafish clinical chemistry blood values, and in **Table 3**, which lists the white blood cell count (Murtha et al. 2003).

Blood Volume

The blood volume in bony fish is small compared to other vertebrates. The total blood volume in fish can range from 1.5 to 3.0% of total body weight, even up to 5% of body weight, depending on references (Conte et al. 1963; Groff and Zinkl 1999). Blood collection in zebrafish is a terminal procedure, and reported blood volume collected from an adult zebrafish ranges from 1 µl to 10 µl (Jagadeeswaran et al. 1999; Murtha et al. 2003). Because of the small sample size, blood samples from different fish are often pooled for analysis. Plasma or serum can be used for biochemical analysis in fish. However, using plasma for clinical chemistry evaluation in fish is preferred since the analyte levels from plasma more accurately reflect circulating analyte levels than from serum (Hrubec and Smith 1999).

Many basic methodologies for common procedures used in other mammalian models have not yet been standardized or developed in zebrafish since the model has been developed primarily to study genetics and gene function in embryos and larvae. However, significant work has been done analyzing blood coagulation, hematology, and hematopoiesis in zebrafish (Jagadeeswaran et al. 1999; Paw and Zon 2000; Bennett et al. 2001; Lieschke et al. 2001; Crowhurst et al. 2002; Gerhard et al. 2002; Onnebo et al. 2004). Groff and Zinkl wrote a review of hematology and clinical chemistry of cyprinid fish focusing on koi, carp, and goldfish. Since zebrafish are also cyprinids, the patterns described for related fishes in this review are probably applicable to some degree (Groff and Zinkl 1999).

Hematology

The most common granulocyte in the zebrafish is the neutrophil (**Figure 11**), which is analogous to the mammalian neutrophil. The

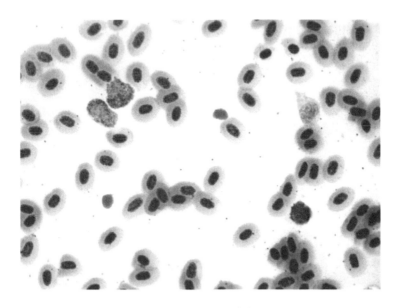

Fig. 11 Neutrophils and nucleated red blood cells from zebrafish blood smear, cytospin preparation. Neutrophils stain positive with Sudan black stain. Oil immersion 1000X. Photo courtesy of Dr. Jan Spitsbergen, Department of Microbiology, Oregon State University.

terminology is not standardized, and the terms *neutrophil* and *heterophil* are both used in the zebrafish literature. However, neutrophil is the most commonly used term. The zebrafish neutrophil has a segmented nucleus with 2 or 3 lobes, while the mammalian neutrophil has 4 or 5 lobes (Onnebo et al. 2004; Lieschke et al. 2001). Zebrafish neutrophils have two types of granules; they are positive for myeloperoxidase and acid phosphatase activity but are negative for periodic acid Schiff (Onnebo et al. 2004; Lieschke et al. 2001; Bennett 2001). The second most common granulocytes are basophils and eosinophils (Onnebo et al. 2004; Lieschke et al. 2001; Bennett 2001). The eosinophilic granules in zebrafish, the basophils and eosinophils, are peroxidase negative, which is different from the human eosinophils (Onnebo et al. 2004; Lieschke et al. 2001). The zebrafish basophils and eosinophils have peripheral, eccentrically placed, unsegmented nuclei that are negative for myeloperoxidase and acid phosphatase activity but that are periodic acid Schiff-positive (Onnebo et al. 2004). Zebrafish macrophages are large, round cells with many cytoplasmic phagosomes (Onnebo et al. 2004). They are negative for myeloperoxidase activity and show weak nonspecific esterase activity (Onnebo et al. 2004; Crowhurst 2002).

2

husbandry

nutrition and feeding

Along with water quality, nutrition and feeding are the most important determinants of success—or failure—in zebrafish facilities. Therefore, to ensure efficient and scientifically sound management of fish stocks, it is essential that managers and technicians possess a thorough understanding of fish nutrition and the types of feeds available, as well as the techniques to deliver them. The following section is an overview of these subjects as they pertain to the culture of zebrafish.

Nutrient Requirements

The ability to rear and maintain fish in captivity is absolutely dependent upon sound nutrition. The specific nutritional requirements of the species being cultured must be determined, and that information should be applied to the development of feeding protocols that promote maximal growth, survival, reproduction, and immune function of stocks.

While the nutritional requirements of zebrafish are unknown, they are likely to be at least to some degree similar to those of warmwater, omnivorous cyprinids, such as goldfish (*Carrasius auratus*), carp (*Cyprinus carpio*), golden shiner (*Notemigonus crysoleucas*), and fathead minnow (*Pimephales promelas*). Data for these species are available and can be used as a reasonable standard when specific information on zebrafish is lacking.

The following is a cursory description of the five major nutrient classes in fish diets, with special reference to data collected on zebrafish, when available.

Proteins/amino acids

Proteins are required in fish diets to provide indispensable amino acids, as well as non-indispensable amino acids or the nitrogen required to synthesize them. Indispensable amino acids are those that cannot be produced by fish and must be supplied in the diet. Amino acids are utilized in a wide number of biological functions, including growth, reproduction, general maintenance, and tissue repair.

The specific demand for proteins in the diet varies considerably both among and within species, and is based on feeding habits in the wild, environmental conditions, and the developmental stage of the animal. Protein requirements, as a proportion of the diet, typically decrease with maturity (Dabrowski 1986).

The total dietary protein requirement for a species can be defined as the minimum amount of protein in the diet necessary to satisfy amino acid requirements and promote maximal growth and reproduction. Dietary protein levels in excess of this amount are metabolized for energy or converted into energy storage products.

Since fish eat to satisfy their energy demands, the balance of proteins and energy in the diet is consequently of critical importance. Diets containing excess proteins will increase ammonia production, which can decrease growth and feed conversion. Diets with excess energy (in the form of lipids or carbohydrates) can decrease food consumption, which will also decrease growth.

The dietary protein and indispensable amino acids for zebrafish are unknown. Minimum protein requirements for other cyprinids vary from 29% to 60% protein as a total fraction of the diet, a range that also appears to adequately support zebrafish growth and reproduction, depending on the setting. In general, levels of crude protein in the diet should be highest for larval fish, and should decrease as the fish enter adulthood. While this holds true for zebrafish, the relative decrease in crude protein requirements as fish age is less than it might be for most other aquaculture species, given the small body size of adults and relatively high maintenance temperatures adhered to in research settings. Therefore, it is acceptable for adult zebrafish diets to contain up to 60% protein, especially for stocks of fish that are bred intensively to meet the demand of large-scale experiments.

Lipids

Dietary lipids are an important source of energy as well as the essential fatty acids (EFA) required for normal growth and development. They also aid in the absorption of fat-soluble vitamins. The polyunsaturated fatty acids (PUFA) linoleate (18:2n-6) and linolenate (18:3n-3) are considered essential because they cannot be synthesized by fish and therefore must be provided in the diet. Most freshwater fish, including zebrafish, have the ability to elongate and convert these PUFAs into the physiologically more important highly unsaturated fatty acids (HUFA), most notably eicosapentaenoic acid (20:5n-3; EPA), docsahexaenoic acid (22:6n-3; DHA), and arachidonic acid (20:4n-6; AA).

Species-specific essential fatty acid requirements vary widely, but species can generally be classified by their relative demand for n-3 and n-6 fatty acids in the diet. Some species require a higher ratio of n-3 FA, some require equal amounts of both, and others require higher proportions of n-6 FA. Zebrafish have been assigned to the last category, as both growth and fertilization have been shown to improve as the ratio of n3:n6 in the diet decreases (Meinelt et al. 1999, 2000).

While the dietary lipid requirement for zebrafish has not been determined, other cyprinids, including carp, golden shiner, and goldfish, all perform well at lipid levels of 10–15% of the total diet. Given this, zebrafish diets should also be within this range, and additionally should contain adequate levels of n3 and n6 PUFA, as well as the HUFAs EPA, DHA, and AA, in order to promote maximum growth and high quality gamete production of stocks (Jaya-Ram et al. 2008).

Carbohydrates

No dietary requirement for carbohydrates has been demonstrated in fish. However, carbohydrates are an important source of energy, and to a varying extent (depending on the species) can be used to "spare" the use of proteins and lipids for that end. Carbohydrates are typically added to prepared fish feeds for this reason, as well as for their binding properties.

All fish analyzed to date, including zebrafish, possess the enzymatic apparatus for the digestion of simple and more complex carbohydrates. However, the ability to digest carbohydrates differs widely among species. To a large extent, this depends upon trophic level; herbivorous fish are typically better adapted for digesting complex carbohydrates than are carnivorous fish.

Zebrafish, as warm-water omnivores that regularly consume aquatic and terrestrial insects (and their chitinous exoskeletons) as well as phytoplankton (Spence et al. 2007), are able to digest and utilize these molecules with some efficiency. In fact, it has been demonstrated that zebrafish display favorable rates of growth, feed conversion, and weight gain when fed diets containing up to 25% total carbohydrate from the larval stage to sexual maturity (Robison et al. 2008).

Vitamins

Vitamins are dietary essential organic compounds required in only very small amounts by fish. To date, 4 fat-soluble (A, D, E, and K) and 11 water-soluble (C, B6, B12, thiamin, folate, biotin, niacin, pantothenic acid, riboflavin, choline, and myo-inositol) vitamins have been shown to be essential in fish diets.

Vitamins must be provided in the diet, although in some instances at least some of these compounds may be produced by metabolism of bacterial flora in the gut. Vitamin deficiencies are notoriously difficult to diagnose because the symptoms are largely non-specific and often develop slowly over time, and the methods for determining precise levels in the diet are complicated and tedious to perform.

Precise requirements for zebrafish are unknown, although certain vitamins, such as ascorbic acid (vitamin C) and retinoic acid (vitamin A), are known to be important for gamete production (Dabrowski and Ciereszko 2001; Alsop et al. 2008). Most live foods are rich in vitamins, and well-formulated prepared feeds should also contain adequate and stabilized levels of these compounds. Together, these feeds should adequately meet the requirements for zebrafish.

Minerals

Minerals are inorganic elements required by fish in trace amounts for a number of biological processes, including ossification, osmoregulation, and nervous system function, and are obtained via the diet and the external aquatic environment. Calcium (Ca), magnesium (Mg), sodium (Na), potassium (K), iron (Fe), zinc (Zn), copper (Cu), and selenium (Se) are all typically absorbed from the surrounding water through the gills. The remaining minerals must be provided in the diet.

Zebrafish are classified as a "hard-water" species, although evidence documenting this is scant. However, their performance is decreased in low Ca^{2+} environments (Chen et al. 2003), suggesting that an elevated metabolic cost may be incurred on fish maintained in soft water (<50 mg/L $CaCO_3$).

Dietary mineral requirements for zebrafish have not been determined; however, diets containing live and/or formulated feeds will
likely provide levels of minerals necessary for biological function and
maintenance. It should be noted that the dietary requirements vary
depending on the concentrations of specific minerals in the water.

Feed Types

There are two general classes of feeds for fish: **live diets** and **processed
diets**. Live diets include a number of zooplankton species, including
Artemia, rotifers (typically *Brachionus* spp.), and *Paramecium*. These
prey items all share a number of favorable characteristics, including
amenability to mass-culture, balanced nutritional profiles, digestibility, and high rates of attractiveness and acceptability. There are
some drawbacks related to their use in research settings, however,
and some of these are discussed in Chapter 5. Processed diets are
designed to replace live diets, and are formulated using biological
materials. The utilization of prepared feeds typically represents a cost
savings over live feeds, allows for greater control over the nutritional
state of research animals, and reduces the overall risk of introducing
pathogens or toxins via the diet. However, the complete replacement
of live prey items with a processed diet is often difficult to achieve,
because the exact nutritional requirements of many species—including zebrafish—are not known. This information, along with the results
of rigorous and extensive dietary studies, needs to be incorporated
into the formulation of these diets to ensure that they most closely
meet the demands of the species in question.

Live feeds

Artemia (also known as **brine shrimp)** nauplii are the predominant
live prey item utilized in zebrafish diets **(Figure 12)**. Aquaculturists
have long exploited an aspect of the reproductive biology of this small
crustacean that results in the production of cysts, which are meta-
bolically inactive eggs encased in a hard shell that, when vacuum
packed and kept in dry and cool conditions, may be stored for years
until use. Upon rehydration in fresh seawater for 18–24 hours, cysts
hatch into free-swimming nauplii that can be fed to zebrafish of all
life stages. Because the nauplii do not feed, subsisting completely off
yolk-sac reserves, it is best to present them to the fish as shortly after
hatching as possible, since the amount of energy transferred from
the nauplii to the fish decreases with nauplii age. Further, this life

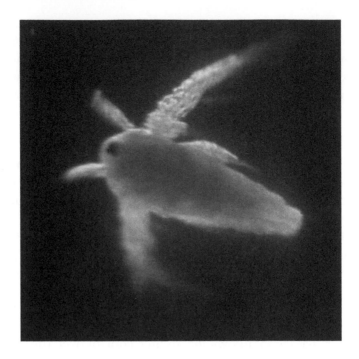

Fig. 12 *Artemia.* Newly hatched *Artemia* nauplii swimming.

stage is the most appropriate for zebrafish in terms of size (~475–500 μm), visibility, and ease of digestibility.

Artemia molt into second-instar metanauplii at approximately 32 hours post-hatch, depending on environmental conditions. These animals are larger (~550 μm), faster, and generally less appropriate for zebrafish, especially larval stages with a smaller gape size and reduced swimming ability. This second-instar stage is also coincident with the onset of exogenous feeding, so if the animals are not provided with a food source from this point on, they will be nutritionally "empty" when presented to the fish. However, as non-selective filter feeders, metanauplii can be "enriched" or "bioencapsulated" with fatty acid emulsifications or microalgae before being fed to fish. Minimum time periods for this process, under "normal" conditions (i.e., incubation in seawater at 24–28°C), are approximately 40–48 hours post-hatch. *Artemia* from the second-instar stage onward can be enriched in this fashion, and are appropriate, if not desirable, for juvenile and adult zebrafish.

Artemia nauplii generally possess a well-rounded nutritional profile, although there may be some variation between strains (or even from batch to batch) in the levels of free amino acids, HUFAs (especially DHA), and biologically available vitamin C. Enriching metanauplii before presentation to fish and/or co-feeding a formulated diet containing some or all of these nutrients can overcome this potential problem.

Live feeds, including *Artemia*, can be contaminated with pathogens (Hoj et al. 2009). For this reason, prior to feeding, cysts should be decapsulated by brief exposure to a hypochlorite/sodium hydroxide solution **(Appendix – Sample decapsulation protocol)**. This process dissolves the hardened outer shell of the animal, thereby greatly reducing a primary source of pathogen transmission. Further, nauplii treated in this fashion have a higher energy content than do non-decapsulated cysts, because removal of the shell shortens the amount of time and energy required for the animal to hatch into a free-swimming nauplius. The energy gained as result of the process can then be transferred directly to the fish.

The two other commonly live prey items most typically used for zebrafish are smaller zooplankton applied during the first-feeding and early larval stages. The first, **Paramecium** spp., is a small (150–250 μm) freshwater ciliate that is often used as a first food source for first-feeding fish larvae, including zebrafish **(Figure 13)**. These animals are simple to mass culture in "semi-sterile" conditions, slow moving (and thus easy to capture), and will continue to live and reproduce in culturing tanks if they are not eaten. Their nutritional value is limited, so they are not suitable for zebrafish beyond the early (first-feeding) larval stage. Their nutritional profile may presumably be

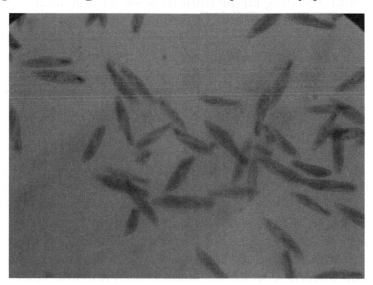

Fig. 13 *Paramecia.* Group of *Paramecium* swimming.

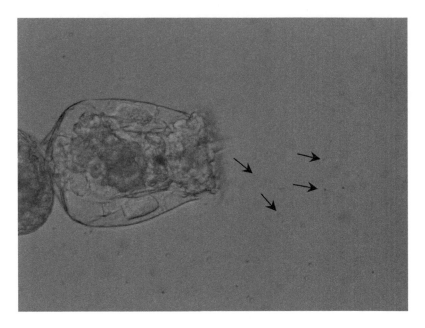

Fig. 14 Rotifer *(Brachionus plicatilis)*. Rotifer is swimming and actively feeding on microalgal cells (black arrows).

improved by feeding them microalgae or *Spirulina* before presenting them to larvae, but their primary value as a prey item is to support growth of first-feeding larvae until their swimming ability and gape size increases such that more nutritionally complete *Artemia* can be consumed with greater efficiency. *Paramecium* is most often the primary component in **infusoria** cultures used in the aquarium trade for the rearing of larval fish.

The third live prey item that is often presented to first-feeding zebrafish is the **rotifer**, *Brachionus* spp. **(Figure 14)**. Rotifers are probably the most widely used zooplankters in all aquaculture for larval fish diets after *Artemia*. They are small (~100–250 μm) and slow swimming, and will readily ingest microalgae, bacteria, *Spirulina*, and fatty acid emulsifications that can then in turn be delivered to fish that feed upon them. The saltwater rotifer *B. plicatilis*, which is the most widely used species in fish culture, is highly tolerant of a wide range of salinities, and will stay alive and reproduce in the lower salinity conditions that are also conducive to larval zebrafish. This fact, along with their small size and elevated nutritive value relative to *Paramecium*, makes them a particularly attractive, if underutilized, prey item for zebrafish larvae.

Processed feeds

Larval Diets: In aquaculture, it is notoriously difficult to achieve satisfactory survival and growth rates in first-feeding larvae using processed feeds. The reasons for this are many, but primarily center around the fact that live prey items are typically more nutritionally balanced, visually and chemically attractive, digestible, and stable in the water column than are processed feeds. A processed diet must possess adequate qualities in all of these areas, and must also demonstrate good water stability (nutrients must not rapidly leach out into the surrounding environment). This is especially important because microparticles must also have the ability to stay in suspension for prolonged periods so that they are constantly available to the fish.

Zebrafish larvae can be reared exclusively on processed diets, but performance is reduced when compared with rearing with live feeds. Reported survival rates can be up to ~80%, and in at least in one instance exceeded 90% (Carvalho et al. 2006). However, some of the best survival rates were obtained under conditions not typically encountered in zebrafish culturing facilities (constant presentation via automatic feeders, large volume tanks, etc.), and in all instances growth was significantly reduced when compared to treatment groups given a live diet.

It is therefore possible to utilize this approach to rear zebrafish, but live feeds are preferable. If a processed diet is utilized, great care should be taken with regard to feed selection and application. Because of their tendency to foul the water, prepared diets should not be overfed, and adequate flow to tanks must be maintained at all times.

Juvenile/Adult Diets: Processed feeds can be utilized as an exclusive food source for cultured fish beyond the larval stage if they are nutritionally balanced, palatable, digestible, water-stable, and result in performance comparable to what would be experienced using a diet comprised of live prey items. However, these standards are not easily met. At minimum, the nutritional requirements of the species for which the diet is being designed must be known, and considerable and rigorous testing must be conducted to evaluate and determine the suitability of the feed for maintaining satisfactory growth and reproduction under culturing conditions.

Until recently, very little work of this type had been done for zebrafish. The nutritional requirements of the species are still not well known, and documented studies testing the efficacy of a prepared feed for promoting performance of zebrafish in culture are limited. However, efforts are currently being made to develop an open-source, standardized diet for zebrafish, using semi-purified, chemically defined ingredients. Preliminary results are encouraging, with test fish fed on first phase standard diets showing enhanced growth when compared with fish fed on various commercially available diets (Siccardi et al. 2009). Still, until these diets move beyond the testing stage, it is not advisable to maintain zebrafish exclusively on these or any other processed feeds.

However, commercially available processed feeds may be co-fed with *Artemia* with excellent results **(Figure 15)**. The advantage to employing this method is that a properly selected prepared feed may provide elevated levels of certain PUFAs, HUFAs, or vitamins that *Artemia* alone may not. Selection of the proper feed is important in this regard. Before widespread application, a new diet should be tested on a small scale. Characteristics to consider include palatability, particle size, suspension in the water column, digestibility, and the resultant biological parameters of growth, fecundity, fertility, and disease resistance.

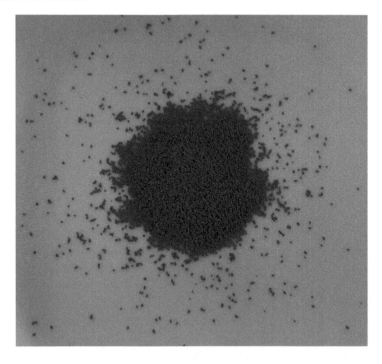

Fig. 15 Typical juvenile/adult processed feed. This is an example of a 400–600-μm pellet that can be added to the diet to provide known quantities of certain key nutrients.

Finally, it is absolutely essential that processed diets (larval included) be stored and administered properly. The typical shelf life of a prepared feed does not exceed 3 months, when maintained in cool, dry conditions (Craig and Helfrich 2002). Oxidation of feed components, particularly fatty acids, increases with temperature. Thus, feeds should be kept in sealed containers, refrigerated, and discarded after 3 months to ensure that fish stocks derive maximal benefit from their application.

Processed feeds should be fed dry, and not hydrated in water and fed via a squirt bottle. Upon exposure to water, water-soluble nutrients in feed particles leach out into the surrounding environment within minutes, resulting in the delivery of nutritionally empty feeds to fish and increased algal growth, among other problems.

If situations dictate that food *must* be hydrated to facilitate feeding (e.g., in facilities with large numbers of tanks), it should be done in such a way that minimizes the amount of time the food is in suspension before delivery. Some microparticulate diets, such as those commonly utilized for larval stages of zebrafish, are encapsulated in protein- or gelatin-based capsules designed to retain water-soluble nutrients for extended periods after exposure to water **(Figure 16)**. Feeds generated in such fashion may be administered to fish in liquefied form with no detrimental effects.

The proper administration of feeds is a major operational challenge in zebrafish culture, requiring a large investment of time and labor, even in the smallest of facilities. Not surprisingly, quality,

Fig. 16 Typical larval-processed feed. This neutrally buoyant, microencapsulated diet has an extremely small particle size of 10–20 μm that is within the gape size of larval fish.

consistency, and ergonomic concerns often arise as a result of having to feed hundreds, if not thousands of individual tanks by hand several times a day. There are two ways to meet these challenges. The simplest and most widely employed response is the development and implementation of standard operating procedures for feed handling and delivery that all care personnel must follow. An alternative solution that has only very recently become an option is the automation of feed delivery to tanks. Commercial automated feeding systems for certain system types are now available, and represent a new and promising way for managers to meet the nutritional demands of fish stocks in a more time- and cost-efficient manner **(Figure 17)**.

Life Stages

As zebrafish develop from their first-feeding, swim-up larvae into reproductively active adults, their nutritional requirements change accordingly, both in terms of quality (nutritional profile) and quantity (density of feeds and frequency of feeding). It is important to understand these changes and to match feeding protocols according to life stage and application of fish stocks.

Larvae

Under typical culturing conditions, zebrafish larvae begin exogenous feeding at approximately 5 days postfertilization. This time period is coincident with a number of developmental milestones, including the completion of a rudimentary digestive system and the opening of the mouth, swim-bladder inflation (and resultant ability to maintain position in the water column), and near depletion of yolk-sac reserves. Over the next several weeks of their lives, until they reach metamorphosis (transition to the juvenile stage, at ~21–28 days postfertilization), larvae must feed nearly constantly to fuel growth and meet energy demands, which are higher at this life stage than at any other point during development.

Given this, there are a host of issues that must be addressed when designing feeding protocols for zebrafish larvae. These include attributes of the food itself—nutrient profile, water stability, suspension, size, digestibility, acceptability—as well as its delivery, including density and frequency of application.

Nutrient Profile: Larval zebrafish require diets rich in protein to help meet growth and energy requirements, which are highest early in development. Diets should also contain adequate proportions of

Fig. 17 Automatic feeding system for zebrafish racks. This robotic feeder delivers measured quantities of live or processed feeds to fish in housing tanks of various sizes. Photo courtesy of Tecniplast Aquatic Solutions.

lipids, in part to help "spare" proteins for growth. Carbohydrates are not required, and should be minimal to ensure that digestion is most efficient. Vitamins must be present in the diet in trace amounts, as should minerals, although these may be obtained from the water. Live feeds are an excellent food source for larval zebrafish because they generally possess well-rounded nutrient profiles and plentiful amounts of free amino acids. Processed diets may also be suitable, given the fact that they have been shown to support growth and survival of zebrafish in some instances (Onal and Langdon 2000).

In general, larval zebrafish should perform well on diets containing up to 45–60% crude protein, 6–10% crude fat, and less than 5% carbohydrate (fiber)(NRC 1993). Trace amounts of required minerals and vitamins should be included.

Stability: Food items must maintain their stability in water such that they do not cause problems associated with fouling, and that their nutritive value is not lost before ingestion by fish. Live diets are obviously very stable, and as long as they stay alive in culturing tanks, adequately retain their nutritive value to fish. Stability is a major problem for processed diets, however, and must be addressed in formulation. Retention of water-soluble nutrients in microparticulate diets is maximized by binding particles in a water-stable matrix or encapsulating them with a cross-linking agent. When utilizing such diets for zebrafish larvae, care must be taken to choose products that demonstrate adequate water stability while retaining digestibility.

Suspension: Larval zebrafish are primarily water column feeders and do not generally possess the swimming ability to efficiently feed on the surface or the bottom. Therefore, feed items must be in suspension in order to be available for the fish. Live diets, such as *Paramecium*, rotifers, and *Artemia*, all swim actively in the water column, and so are generally well suited for feeding to zebrafish larvae. However, *Artemia* will typically die within several hours after immersion in fresh water, and so will not be available for extended periods. Slight aeration in tanks will aid in prolonging this, but once the animals expire, they are of no value to larval fish. Processed diets must be of low enough density to ensure they remain in the water column for adequate periods. This too can be aided with minimal aeration/circulation. Diets that rapidly sink to the bottom are of little value, especially early on, and can cause problems associated with the fouling of water in culturing tanks. The requirement for food to remain in suspension becomes less absolute as fish grow.

Size: Size of the prey item is also crucial. Larval fish are gape-limited predators that normally ingest their prey whole. Consequently, prey items must be of the appropriate size for the life stage. A prey size of 150–200 µm is the maximal size range acceptable for first-feeding zebrafish **(Figure 18)**. *Paramecium*, rotifers, and a variety of processed diets are appropriate in this regard. Freshly hatched *Artemia* nauplii are typically a bit large for first-feeding zebrafish (500–550 µm in length), but the width (<200 µm) is within range, and it is

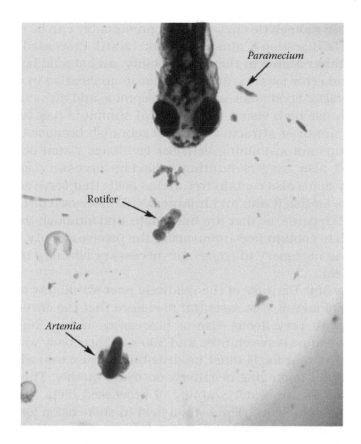

Fig. 18 First feeding zebrafish larvae, *Paramecium, Brachionus* roti-fer, and *Artemia* nauplii. *Paramecia* and rotifers are both well within the gape size of first-feeding zebrafish, while *Artemia* nauplii are larger and more difficult for them to handle.

possible (although more difficult) to rear zebrafish on them exclu-sively without an intermediate step (Carvalho et al. 2006). Processed feeds are typically available in a wide variety of different-sized par-ticles, and should be selected appropriately. An initial particle size of 50–100 μm is suitable for zebrafish from the larval stage up to metamorphosis; they will consume higher numbers of particles per feeding as they grow.

Digestibility: Because larval zebrafish have simple diges-tive systems, they have limited capacity for nutrient absorption. Consequently, feeds must be highly digestible, such that larvae can readily absorb nutrients and assimilate them for growth and metab-olism. Live prey items are typically highly digestible, in part because they contain proteolytic enzymes that presumably can be utilized by fish to aid in digestion (Cahu and Infante 2001). Processed feeds will vary to a large degree in their digestibility, and should be analyzed by trial and error for this attribute prior to application in diets.

Acceptability: Food items must be acceptable and attractive to fish. Larval fish use both visual and chemical stimuli to help locate food. Live feeds are most attractive to larval zebrafish because their slow-moving, constant swimming behavior facilitates visual detection by larvae and also triggers hunting behaviors necessary for feeding. These prey items also contain free amino acids that serve as olfactory attractants for fish (Cahu and Infante 2001). Processed diets are inferior in both regards, as they are inanimate, and although they may be formulated to contain free amino acids, the precise identity and quantity of those necessary to trigger the necessary olfactory response is often unclear.

Feed Density: Because of the relatively poor swimming ability and small size of larvae, it is essential to ensure that the encounter rate of larvae with prey items—live or processed—is very high. Larval fish can consume between 50% and 300% of their body weight daily, which means that foods must be distributed to excess, while at the same time not impinging negatively on water quality. This also has implications for the water stability of processed diets, which must retain water-soluble nutrients when held in suspension for extended periods of time. Live feeds, such as *Paramecium* and rotifers, are excellent in this regard because they can be added to culture tanks at very high densities without any detrimental effects on the fish. They also will survive and in some instances reproduce in tanks, serving as a constant supply of food for developing fish.

Processed feeds may also be applied at high densities; however, they should be applied frequently and in small doses to guard against fouling. Water stability of such feeds is essential; encapsulated feeds that retain water-soluble nutrients when in suspension for extended periods are greatly preferred in this regard because they are less likely to quickly foul the water and they will provide greater nutrition for larvae that consume them.

Feeding Frequency: Because of their high metabolic rates, larval zebrafish must be fed intensively. Frequent, small meals of live and or processed feeds will support the best growth rates. Adding high densities of live prey items that can survive for extended periods in tanks is the best way to achieve this, as larvae exposed to high densities of zooplankton will have a constant source of high quality food available to them. This approach can also be implemented by using processed feeds that demonstrate good water stability and suspension, but is more labor intensive and less practical when automatic feeders that continuously dose small amounts of feed cannot be used.

Juveniles

Once larval zebrafish have metamorphed into juveniles (~21–28 days postfertilization, depending on conditions), larval tissue, organs, and structures have been replaced by immature adult versions of the same, and the animals take on the appearance of miniature adults. This maturational phase of zebrafish development is still characterized by rapid growth, and the diet must reflect this demand.

Nutrient Profile: As in the larval stage, protein requirements are still very high, given the rapid rates of growth that must be attained as the animals mature into reproductively active adults. Lipids are still very important, both in terms of the protein-sparing effect of lipids as an energy source, as well as for sexual maturation. Carbohydrates too may also be used to spare proteins for growth to a greater extent, as the digestive system is more developed, although there is no requirement for this. Minerals and vitamins are also of the same importance that they were during the larval stage to help support growth and metabolism during this phase.

In general, diets used for juveniles should have similar profiles to larval diets: 45–60% crude protein, 6–15% crude fat, less than 5% carbohydrates (NRC 1993), and trace amounts of necessary vitamins and minerals.

Stability: It is still important that feeds maintain their stability in water during the juvenile phase. The salient difference during this stage of development is that this is the point at which the fish may be weaned onto pelleted feeds, which typically are much less stable in water than are microparticulate, encapsulated diets used for first-feeding larvae. Appropriate care must be taken to ensure that such feeds are not overfed and are delivered properly. This is particularly important since it will take several days for fish to become accustomed to feeding on a new feed type.

Suspension: Because juveniles are much more accomplished swimmers and hunters than they are during the larval stage, they become less restricted to the water column in terms of feeding. They will readily learn, over the course of a few days, how to take food at the surface, and are more adept at picking particles off the bottom of tanks. Still, feeds that remain in suspension for prolonged periods are desirable, both because feeding efficiency is still greatest in this area, and because a greater proportion of feeds that sink to the bottom are flushed out of tanks and into sumps before can be eaten.

Size: Gape size in juveniles is increased, and while they still consume prey items whole, they are more adept at tearing and breaking off pieces of feeds when necessary. Therefore, the size of feed particles that they can ingest is increased, such that they readily ingest prey in a size range of 400–600 μm, which includes *Artemia* up to the second-instar metanauplius stage. Processed feeds within this range are readily available in crumble or pellet form.

Digestibility: Although juveniles possess better developed and more complete digestive systems than do larval fish, feeds should still be highly digestible to help minimize energy loss. Energy "gained" in this manner can be allocated to growth and maturation, rather than to the breakdown of complex molecules.

Acceptability: The same principles regarding food acceptability and palatability that apply to larval zebrafish also apply to juveniles. Juvenile zebrafish are probably less selective, as evidenced by their willingness to feed on a greater variety and presentation of processed feeds, although this may be more indicative of enhanced swimming ability and increased gape size than it is of visual or olfactory preferences.

Feeding Densities: Because juveniles are still in a rapid growth phase, they should still be fed to excess (up to 25–50% body weight daily)(NRC 1993). It is less feasible, at least under typical culturing conditions, to keep a constant supply of zooplankton growing in the tanks along with the fish, primarily because *Paramecium* are no longer nutritious or large enough to be a major portion of the diet, and the most important zooplankton food source at this point, *Artemia*, will not live in culture tanks much beyond a few hours. Further, juvenile fish are also much more efficient feeders, and typically will consume much larger amounts of prey during feedings. Therefore, when using live diets—at this point primarily *Artemia* (although rotifers are still feasible)—it is necessary to shift the feeding approach more toward administering frequent, small meals. This approach can also be employed when using processed feeds, and because flow rates can be increased as juvenile swimming ability improves, fouling is not as big a threat as it is under static or low flow larval rearing conditions. Still, overfeeding should be avoided, and feeding the fish frequent, smaller meals that result in little or no uneaten food will facilitate this.

Feeding Frequencies: The frequency with which food should be presented is strongly correlated with gastric evacuation time (how long it takes the gut to empty once full). This is particularly relevant in juveniles in the rapid growth phase. These data are not available for zebrafish. However, since zebrafish do not possess a stomach, it is likely that residence time in the intestine determines the assimilation efficiency. Therefore, if fish are fed too frequently, residence time, and therefore assimilation, is reduced. If they are fed too infrequently, demands for energy and growth are not met. Evacuation times are rapid in juveniles, resulting from their small body size, so feeds must be presented frequently in small portions, throughout the daylight hours. Ideally, they should be fed 4–8 times daily. Having constant sources of live zooplankton, such as rotifers, or even *Paramecium* (to a lesser degree), in culturing tanks can improve situations in which this frequency cannot be maintained owing to logistical constraints associated with repeated hand feeding.

Adults

Zebrafish are primarily reared for gamete production. Upon attainment of sexual maturity, zebrafish stocks should be able to produce large numbers of embryos on a consistent, if not continuous, basis until they become senescent or expire. The diet should be designed to support this goal.

Nutrient profile: Although adult zebrafish continue to grow throughout their lifetime, they are no longer in a rapid growth phase. Therefore, beyond what is needed to maintain homeostasis, the emphasis of the utilization of dietary nutrients shifts from growth to reproduction. Protein should still comprise the major fraction of the diet, but the requirement is reduced compared to the juvenile and especially the larval stage. Lipids become more important, as certain fatty acids (n3 and n6 PUFAs and HUFAs) are important for reproductive function, particularly egg production and fertility (Jaya-Ram et al. 2008). Carbohydrates are not required, and still should be at minimal levels to avoid undue energy expenditure. The dietary requirement for minerals and vitamins is still minimal relative to their proportions of the total diet, although certain vitamins may be required in slightly greater amounts. For instance, elevated levels of the antioxidant vitamin C (ascorbic acid) in the diet have been shown to exert positive effects on gamete production and quality, particularly as fish stocks age (Dabrowski and Ciereszko 2001). Vitamin A has also been shown to be necessary for reproductive function in zebrafish (Alsop et al. 2008).

In general, diets used for adults should have nutrient profiles in the range of 45–55% crude proteins, 10–15% crude fats, less than 5% carbohydrates, and trace amounts of necessary vitamins and minerals (NRC 1993). Feeds that contain sufficient elevated amounts of ascorbic acid, Vitamin A, and others may also help long-term maintenance and reproductive function of stocks.

Stability: Feeds should demonstrate good water stability. This is important because it is desirable to ensure that the nutrients in the food are getting to the fish, and not the system (and algae and cyanobacteria). This is particularly relevant when pellet, crumble, or flake feeds are being co-fed with *Artemia* in an attempt to supplement fatty acids and/or vitamins requirements. These are exactly the nutrients that will leach out of feeds most rapidly upon hydration, so care should be taken to ensure that feeds are distributed properly.

Suspension: Adult zebrafish feed equally well at the surface and bottom and within the water column. However, feeds that spend the greatest time in suspension—or on the surface—are preferable, since zebrafish are adapted to feed at the surface and within the water column, and because a greater proportion of sinking feeds will be flushed from tanks before they can be consumed.

Size: Although adult zebrafish will readily ingest prey items in excess of 600 μm, it is best to keep prey items in the range of 400–600 μm to facilitate digestion. Processed diets that are larger, such as flakes, should be ground down into smaller pieces before being fed. Most pellet and crumble feeds are available in graded sizes and can be selected accordingly.

Digestibility: Feeds should still be highly digestible to help minimize energy loss. Energy "gained" in this manner can be allocated toward reproduction, rather than to the breakdown of complex molecules.

Acceptability: The same standards that apply to larval and juvenile stages also hold for adults, although they are somewhat less selective. However, there will be some processed feeds that are less attractive than others, and this should be taken into account. This is especially important relative to residence time in suspension and water stability.

Feeding Densities: Adult zebrafish are often fed to satiation, which is a somewhat nebulous term. The most appropriate approach is to design feeding protocols in a manner similar to that used in most aquaculture species, which are typically fed between 1% and 5% of their body weight each day (NRC 1993). Adult zebrafish may be fed more than this, up to 10% depending on intensity of use, given the fact that unlike most aquaculture species, they are produced for gamete production and not for flesh. Breeding, especially on a constant basis, is energetically expensive and stressful, and therefore fish need to recoup energy and nutrients lost during the process in the diet. Care should be taken not to overfeed, however, particularly when using processed feeds. This can result in poor water quality, which will negatively impact reproductive function.

Feeding Frequencies: Because of the energy costs associated with gamete production, adult zebrafish in typical culture situations have an elevated nutrient demand and need to be fed intensively in order to maintain high levels of production. As with juveniles, the ideal situation is to feed adults frequent, small meals, such that the feed ingested can be digested properly, and that the interval between meals is not too long. Data on gastric evacuation times in adults arc not available, but it is reasonable to assume that adults should be able to tolerate longer intervals between feedings than juveniles because of their larger body size. Feedings should be distributed throughout the daylight hours, ideally from 3 to 5 times daily (at 5–10% of body weight).

It should be noted that zebrafish not being utilized for gamete production have much lower nutrient demands, and can be fed at greatly reduced densities and frequencies.

breeding

One of the most important features of zebrafish as a model animal is their great fecundity. Zebrafish mature rapidly, will spawn readily under a wide range of conditions, and can produce many thousands of offspring over their relatively short reproductive lifetime. This impressive reproductive potential makes them amenable to a wide range of biological studies, and greatly improves the statistical power of experiments in which they are used.

Reproductive Cycle

Zebrafish typically attain sexual maturity within 3 to 6 months postfertilization in laboratory settings. As this may vary considerably with environmental conditions, including rearing densities, temperature, and food availability (Spence et al. 2008), it is most appropriate to relate reproductive maturity to size rather than age. A standard length of approximately 23 mm corresponds with attainment of reproductive maturity in this species (Eaton and Farley 1974; Spence et al. 2008).

Under favorable conditions, zebrafish spawn continuously upon attainment of sexual maturation (Breder and Rosen 1966). Females are capable of spawning on a daily basis (Eaton and Farley 1974; Spence and Smith 2006), although the length of the interspawning interval is dependent upon nutrition, water quality, and intensity of production (Lawrence 2007). A resting period of 1 week between controlled spawning events (setups) is often recommended for optimal production (Westerfield 2007).

Controlling Factors

Olfactory cues play a determining role in zebrafish reproduction and spawning behavior (**Figure 19**). The release of steroid glucuronides into the water by males induces ovulation in females (Chen and Martinich 1975; Vandenhurk and Lambert 1983), and females exposed to those pheromones show significant increases in spawning frequencies, clutch size, and egg viability when compared with females held in isolation (Gerlach 2006). Upon

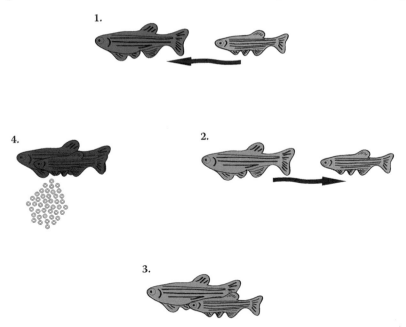

Fig. 19 Olfactory control of zebrafish spawning. (1.) Male zebrafish (orange) releases pheromone that triggers ovulation in the female (blue). (2.) The "ovulated" female (green) then releases a pheromone that triggers courtship and chasing behavior in male. (3.) The "activated" male (green) then courts and chases the ovulated female. (4.) Oviposition: female (red) releases eggs and male (red) fertilizes them.

ovulation, females release pheromones that in turn prompt male mating behavior, which immediately precedes and elicits oviposition and spawning (Vandenhurk and Lambert 1983). Given the positive impact these interactions are likely to have on breeding efficiency, it is important to keep males and females together in the same water for at least 12 hours before setting them up in crossing situations.

Pheromonal release also may suppress reproduction, as holding water from "dominant" female zebrafish has been shown to inhibit spawning of subordinate females (Gerlach 2006). Therefore, it is advisable to avoid holding females in the same housing tank with one another for extended periods of time to prevent such counterproductive dominance hierarchies from being established.

Reproduction in zebrafish is also strongly influenced by photoperiod. Ovulation most typically occurs just prior to dawn (Selman et al. 1993), and spawning commences within the first few hours of daylight (Spence et al. 2006; Engeszer et al. 2007). However, breeding is not strictly limited to this time period. Zebrafish will spawn throughout the day, particularly during the evenings, although egg production is most reliable and intense in the early morning. In the wild, zebrafish have also been observed spawning during the afternoon following the onset of heavy rain (Spence et al. 2008). This "flooding" effect may be readily replicated in the laboratory by flushing crossing cages with fresh water.

Reproductive Behavior

Zebrafish display ritualized courtship behaviors prior to and during spawning. During courtship, males swim in tight circles or hover, with fins raised, above a spawning site in clear view of nearby females. If females do not approach, males will chase them to the site, snout to flank. When spawning, a male swims parallel to a female and wraps his body around hers, triggering oviposition and releasing sperm simultaneously (Spence et al. 2008). Males compete for access to females, and will establish and defend territories (Darrow and Harris 2004; Spence and Smith 2005). These observations, along with the fact that females will produce larger clutches and spawn more frequently when paired with certain males (Spence and Smith 2006), indicate that females are selective.

Females choose mates upon the basis of several factors. The quality of a spawning site is clearly important, as both male and female zebrafish show a strong preference for the oviposition site, selecting

Fig. 20 Typical zebrafish static mating cage set up to promote breeding behavior. Note that plastic plants (arrow) have been added to the tank and the insert has been slanted to provide a shallow spawning area (arrow).

and spawning over gravel versus silt in both laboratory and field-based experiments (Spence et al. 2007). If given the choice, fish will also spawn preferentially in vegetated versus non-vegetated sites (Spence et al. 2007), and in shallow versus deep water (Sessa et al. 2008). These preferences can be facilitated in the laboratory by adding plastic plants to crossing setups, and by presenting fish with shallow water in which to breed **(Figure 20)**.

Females also select males based on their genotype. Zebrafish use olfactory cues to distinguish between kin and non-kin, and this mechanism may be utilized during spawning to avoid inbreeding. Adult female zebrafish prefer the odors of non-related, unfamiliar (reared and maintained separately) males over those of unfamiliar brothers (Gerlach and Lysiak 2006). Obviously these preferences are not absolute, as the fish will mate with siblings/relatives. However, it is important to keep such natural tendencies in mind when designing breeding programs and troubleshooting egg production problems.

Spawning Techniques and Equipment

There are a number of methods that can be employed to breed zebrafish in laboratory settings. These various strategies and affiliated equipment can be generally categorized into one of two basic approaches: "in-tank" and "static-tank."

In-tank strategies

"In-tank" breeding strategies involve the placement of a spawning site or substrate directly inside holding tanks, while fish remain "on system" or in flow. This type of technique relies on the "natural" production of fish kept in mixed sex groups with minimal manipulation of individuals. Another important feature of this basic approach is that because fish remain on flow, water quality is regulated and maintained throughout breeding events. Finally, it also largely minimizes the handling of fish, which can be a stressful event (Davis et al. 2002).

The first formally described technique for breeding laboratory zebrafish is the most basic example of an in-tank breeding method. In this approach, glass marbles are placed at the bottom of holding tanks to provide a spawning substrate for the animals. Fish spawn over the marbles, and the eggs drop into the spaces in between, preventing egg cannibalism and facilitating their subsequent collection by siphoning (Westerfield 2007; Nusslein-Volhard and Dahm 2002). While this method may be effective to some extent, it is generally impractical for use in most facilities, especially those that contain hundreds or thousands of tanks.

A more advanced in-tank approach involves placing a breeding box or container in holding tanks that fish will spawn over during breeding events **(Figure 21)**. A common feature of this method is that the box/container will have a mesh-type top through which spawned

Fig. 21 Breeding box. A breeding box of the type in this photo is placed directly inside holding tanks. Photo courtesy of Dallas Weaver.

eggs drop and are subsequently protected from cannibalism. The box will also typically have some plastic plants affixed to it to make it more attractive as a spawning site. This type of method is more facile than the marbles technique, as boxes can be moved freely in and out of holding tanks as desired. It also better facilitates the collection of staged embryos from groups of fish, and can also be used for breeding pairs. This method is utilized relatively infrequently, and thus no commercially fabricated equipment of this type is available. When this method is chosen, the box/container must be custom-made to fit with the needs of the particular facility in which it is being utilized.

A recent advance in in-tank breeding technology is the Mass Embryo Production System (MEPS™) (Aquatic Habitats, Apopka, Florida, USA). The MEPS is a large spawning vessel, with a holding capacity of 80 or 250 liters that can be operated with its own life support (biological filtration) or plumbed directly into any existing housing system. The unit, which can house large populations (up to 1000 or more) of breeding fish, contains one or more spawning platforms, which are specially fabricated funnels capped with plastic mesh screens that can be located at various depths inside the vessel **(Figure 22)**. When the spawning platforms are placed inside the vessel, fish breed over and on the platforms, and spawned eggs fall through the mesh into the associated funnels. The eggs are then pumped through an attached tube into a separate collection screen by means of pressurized air directed into the funnels, allowing eggs to be collected without disturbing the fish. The units also have the capability to be run on altered photoperiods via the use of an attached light-cycle dome with a programmable light-cycle dimmer.

The MEPS system capitalizes on several attributes of the general in-tank breeding approach, including consistent water quality and minimal handling of animals, with the added benefits of reduced labor input and increased space efficiency. When used properly, this technology is capable of supporting high-level egg production on the order of tens of thousands of embryos per event, and is therefore well suited for experimental applications requiring large numbers of time-staged eggs.

Static-tank strategies

The alternative to in-tank breeding strategies is to remove fish from holding tanks and to spawn them in off-system or "static-water" breeding chambers. This general approach, which is utilized in the great majority of zebrafish breeding facilities, adheres to the following general principles: a small (typically <1 L) plastic mating cage or

Fig. 22 MEPS (Mass Embryo Production System). Large numbers of fish can be housed and spawned directly inside the unit to facilitate on-demand collection of embryos. Photo courtesy of Aquatic Habitats, Inc.

insert with a mesh or grill bottom is placed inside a slightly larger container that is filled with water. Fish (pairs or small groups) are then added to the insert in the evening. When the fish spawn, the fertilized eggs fall through the "floor" of the insert and are thereby protected from cannibalism by adults (Mullins et al. 1994).

This technique has proven to be generally effective and, consequently, derivations of the static mating cage are manufactured by a number of system vendors and laboratory product supply companies.

Fig. 23 Typical static breeding chamber.

Available products vary slightly in size, shape, depth, and total volume, as well as in adjustability of inserts in the static spawning chamber **(Figure 23)**.

The effects of these subtle variations are not well understood, although it has been shown that breeding success is greater when inserts are tilted to provide a shallow to deep gradient inside the chamber (Sessa et al. 2008). This configuration works well presumably because it facilitates the preference of fish to spawn in shallow water (Spence et al. 2008).

There are a number of strengths to the static tank approach. Virtually any type of experiment can be supported using this technique, as fish of any desired genotype can be set up in pairs or smaller groups in a varying number of crosses. Because fish are removed from holding tanks, the effects of behavioral hierarchies established in holding tanks that can be counterproductive to breeding are negated. Static tank technologies also allow for direct manipulation of water quality parameters. Changes in water chemistry, such as decreases in salinity, pH, and temperature, are thought to promote spawning in fish adapted to monsoonal climate regimes (Munro et al. 1990). These factors may also affect reproduction in zebrafish (Breder and Rosen 1966).

There are drawbacks to static-tank breeding strategies. Because the chambers are off-flow, water quality conditions in the spawning setups deteriorate over time. Although this has not been formally investigated, metabolites, particularly total ammonia nitrogen (TAN), accumulate in the water and are likely to have a negative effect on spawning. Tanks may be flushed with fresh water to offset these potential problems, but this represents added labor. Using static set-ups also necessitates that fish are handled constantly, which may be a source of long-term stress for breeding populations. Finally, although it is possible to support experiments requiring large numbers of embryos using current static breeding technologies, it is both labor and space intensive to do so, especially when compared with more advanced in-tank breeding technologies, such as the MEPS.

larviculture

A vital component of any zebrafish research program is the ability to rear stocks of fish from the embryonic stage through adulthood. The efficacy of a rearing protocol can be defined by three major performance factors on the nursery: growth, survival, and labor input. The "ideal" or "optimal" approach is one that requires minimal labor input and results in the highest rates of both growth and survival. The achievement of this goal is a challenging proposition that requires keen observational skills and a thorough understanding of how nutrition and water quality relate to larval biology and behavior.

Embryo Handling and Care

A critically important but often overlooked first step of zebrafish larviculture happens well before the fish are even placed in the nursery. Indeed, the manner in which fertilized embryos are handled after collection in spawning events is a crucial determinant of later success in rearing. After eggs are spawned, they sink to the bottom of breeding tanks, where they will collect with waste materials, including feces, shed scales, and uneaten feed. Embryos should be separated from this organic refuse in relatively short order, as prolonged exposure to these substances will increase their susceptibility to fungal and/or protozoan contamination.

The simplest way to collect eggs is to gently pour the water containing the eggs through a non-abrasive tea strainer **(Figure 24)** or a soft nylon net with a pore diameter of 600 μm or less **(Figure 25)**. The eggs that are retained on the net or screen should then be gently rinsed with clean water before being transferred into standard 90 × 20-mm petri dishes filled with clean embryo medium or fish water (Westerfield 2007; Nusslein-Volhard and Dahm 2002). It is important to limit densities in dishes to 50–100 eggs per dish to help limit fungal

Fig. 24 Tea strainer. Tea strainers such as this can be used to collect zebrafish embryos after spawning.

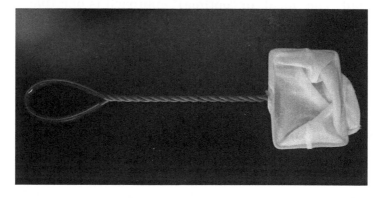

Fig. 25 Soft nylon net. This soft nylon net is designed for the collection of brine shrimp, but can also be used to collect zebrafish embryos after spawning.

and protozoan contamination and reduce developmental asynchrony in clutches (Curado et al. 2008) **(Figure 26)**. In some instances, it may also be beneficial to add 0.00003% methylene blue to the solution to further reduce the likelihood of fungal infestation of the eggs (Nusslein-Volhard and Dahm 2002; Westerfield 2007). Twelve to 24 hours after collection, non-viable embryos should be removed from dishes and the water replaced. The eggs may also be surface-disinfected with a mild bleach solution at this time, in accordance with standard protocols (Westerfield 2007). After this point, the eggs may be left undisturbed in incubators until the larvae hatch and inflate their swim bladders.

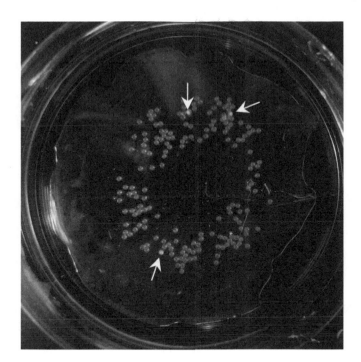

Fig. 26 Petri dish with zebrafish embryos. It is important to limit embryo densities to 50–100 per dish to limit fungal/protozoan contamination and facilitate the removal of non-viable embryos (white arrows).

Larval Biology

Under typical laboratory conditions, zebrafish larvae usually hatch 2.5–3 days postfertilization (dpf) (Westerfield 2007). The timing of this (and all developmental events) depends to a large extent upon temperature; at higher temperatures (28–30°C) it may be a bit sooner, while at lower temperatures (24–26°C) it will be later. Upon hatching, larvae sit on the bottom or sides of tanks and display low levels of spontaneous activity **(Figure 27)** until ~5 dpf, when they inflate their swim bladders by swimming up and swallowing air at the water surface (Goolish et al. 1999) **(Figure 28)**. This allows them to control their buoyancy and regulate their position in the water column, which is a key requirement for feeding and capturing prey.

Up until this point, larvae are completely dependent upon yolk-sac reserves for nutrition. Once exogenous feeding begins, they still derive some benefit from the yolk for a day or two, but the energy required to swim and digest feed quickly depletes the yolk, and it

Fig. 27 Newly hatched zebrafish larvae. These fish have just hatched and have affixed themselves to the sides of tanks by specialized cells on the head (arrows).

Fig. 28 Swim-up zebrafish larvae. These fish have just inflated their swim bladders and now freely swim in the water column, just beneath the surface (arrow).

is completely absorbed by approximately 7 dpf (Jardine and Litvak 2003). For the purposes of larval rearing, it is imperative to begin presenting the fish with feed as soon as they are up in the water column, so that the overlap between first feeding and total yolk sac absorption is maximized. This effectively gives the larvae a nutritional head start during this period. If the fish are not fed at all, they will starve by 10 dpf (Jardine and Litvak 2003).

Once their yolk sacs have been depleted, larval zebrafish need to feed nearly continuously to meet protein demands, which are highest during this stage of development. First-feeding larvae have very simple digestive systems, so feed items must be highly digestible in order for them to maximize nutrient uptake. In nature, zebrafish larvae feed primarily on small zooplankton. Correspondingly, in culture, they are most efficient at capturing items suspended in the water column, although they will develop the ability to also take items at the surface and on the benthos (bottom) as they mature.

Feeding

The feeding of larval zebrafish is discussed at length in the Nutrition and Feeding section of this chapter.

Water Quality

A major challenge in the rearing of any species of fish in captivity is to balance the need for the larvae to feed constantly with the maintenance of suitable water quality in their environment. While zebrafish do not require pristine conditions during the larval stage, large, rapid swings in certain environmental parameters, such as ammonia and especially dissolved oxygen, can quickly kill them. Gradual, unchecked changes in these parameters will also result in mortality if they exceed or drop below certain thresholds. Therefore, the goal should be to maintain stability within a reasonably wide range of conditions.

Larval zebrafish do not produce significant amounts of waste, so the primary concern with respect to the maintenance of water quality in nursery tanks is the fate of uneaten feed items, which will produce ammonia and consume oxygen as they decompose. If these wastes are not removed, water quality will deteriorate to a point where the environment will no longer sustain the fish. The easiest way to offset this problem is to match feed types and amounts with flow regime.

Fig. 29 Static water nursery tank. Nursery tank with static water containing high densities of *Paramecia* and first-feeding zebrafish larvae.

Larval zebrafish can be kept in static water (off-housing systems, no flow) if live diets that persist in rearing water, such as *Paramecium*, are used, or if a percentage of water is changed and uneaten feed is removed on a daily basis **(Figure 29)**. The advantage of keeping larval fish in static water during early larval development is that they will not have to expend excess energy swimming against currents and the depths of water in tanks can be kept shallow to help maximize encounter rates with feed items. However, this approach is generally suitable only for the first week to 10 days post-hatch, as *Paramecium* does not meet the nutritional requirements of the fish once they approach metamorphosis (~15–20 days, depending on conditions) (Lawrence 2007). The application of feeds with nutritional profiles more appropriate for fish that have passed through the first feeding stage, such as *Artemia* (which perishes in fresh water) and various processed diets, requires frequent water changes, which quickly becomes impractical from a labor standpoint. Frequent manual removal and addition of water from tanks can also be viewed as disturbances that may be stressful for the fish.

At this point, or immediately upon first feeding if *Paramecium* is not used, nursery tanks should be placed on housing systems so that they receive flow. The simultaneous addition of clean water and removal of wastewater achieved when tanks are on recirculating systems serves to maintain water quality parameters within the acceptable range. Care should be taken to ensure that flow rates are not too high: feed items may get washed from tanks before larvae have a chance to consume them, and energy expended by the fish to maintain position in the water column and capture prey comes at the expense of growth (Bagatto et al. 2001). Flow should begin at very slow drips (1 to 2 drips every 10 seconds) and increased gradually over time as the fish grow and become more adept at capturing and consuming higher quantities of more complex types of food.

The quantity of food given during each feeding event is also important for maintenance of suitable water quality, especially for items that do not persist in the rearing water. Live items that survive in rearing water, such as *Paramecium*, can be presented in larger, less frequent doses, because they will not foul the water and remain available for the fish to consume on an on-demand basis. Early on, water-stable, neutrally buoyant processed feeds may also be dosed in this fashion because of their persistence in the water column. Perishable feeds, such as *Artemia* nauplii and most processed diets, should be presented in smaller, frequent doses to minimize waste. Larval fish will not eat dead *Artemia* that have sunk to the bottom of tanks.

Finally, the amount of food given at each feeding should be gradually increased in accordance with the amount of food that the fish will actually consume. This can be assessed by direct visual observation of the fish a few minutes after feeding. The "right" feeding amount will result in fish with full guts and little to no uneaten food on the bottom of the tank.

housing density

The density at which fish are kept in captivity exerts profound effects on their health, productivity, and welfare. In research settings, holding densities will also ultimately have logistical implications for experiments relative to costs, space, and labor. In general, keeping fish at higher densities results in lower values for all three of those parameters, while lower densities tend to increase them.

The optimal holding density for a species may be defined as the number of animals that can be maintained in a defined space over the long term while safely promoting their maximal welfare and productivity. The two primary determinants of this value in laboratory conditions are the specific behavioral tendencies of the species in question and the quality of the environment. The equation is further complicated because fish undergo enormous physiological and behavioral changes as they mature, so the optimal density for a larval animal is different from what it is for an adult. Therefore, the criteria for setting maximal *and* minimal density thresholds are extremely complex.

The most evident "readouts" of density are survival, growth, and reproductive rates, all of which should be at their highest when densities are optimal. There are a number of additional, though less obvious, population characteristics that can be used to evaluate the appropriateness of densities, including resistance to disease, the frequency of aggressive interactions in tanks, sex ratios, and circulating levels of stress hormones. Ideally, guidelines for densities should be based on data for all these factors.

For zebrafish, data of this nature are limited. The fact that these animals are a schooling species is well established (Engeszer et al. 2004; Engeszer et al. 2005; Spence et al. 2008), and suggests that it may be acceptable to stock them in higher numbers of fish per unit volume. However, they do not appear to tolerate higher densities, as evidenced by reduced breeding efficiencies (Goolish et al. 1998) and increased stress hormone levels (Ramsay et al. 2006). Further, from a practical management standpoint, as fish densities increase, it becomes more challenging to deliver feed amounts sufficient enough to support growth and reproduction while also maintaining optimal water quality.

At the same time, lower stocking densities may also be problematic, as zebrafish are known to display aggressive behavior associated with the establishment of territories and dominance hierarchies (Larson et al. 2006). Such agonistic interactions tend to increase as density is reduced (McCarthy et al. 1992), and may result in increased energy expenditures on the part of all fish, as well as a chronic stress response in subordinate individuals (Fox et al. 1997). The latter situation can lead to reduced disease resistance in stocks (Harris and Bird 2000).

Densities experienced during the larval and juvenile stages also appear to affect sexual differentiation in zebrafish. In many fish species, the expression of gender may be influenced by a variety of environmental factors during sexual development, including rearing densities. This is the case for zebrafish, as reduced densities promote female differentiation while higher densities result in a higher proportion of males in the population (Nusslein-Volhard and Dahm 2002; Lawrence et al. 2008). Thus, rearing fish at elevated or reduced densities may result in the overproduction of one sex over the other, potentially complicating experiments.

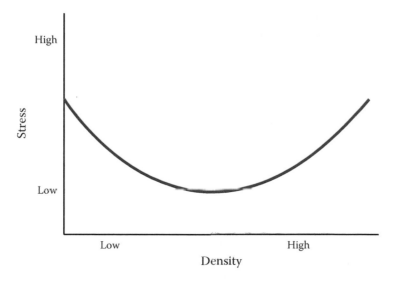

Fig. 30 Model of the effects of housing density on stress. Stress of laboratory zebrafish is increased at both low and high densities because of aggressive interactions and crowding effects, respectively. Stress is lowest at intermediate densities.

That both crowding and low-density situations are less than ideal for zebrafish is very likely because they are a shoaling species, and this highlights the importance of matching holding conditions to behavioral tendencies. A reasonable interpretation of existing data on zebrafish behavior is that housing them at intermediate densities is most conducive to optimal production and welfare **(Figure 30)**; however, this generalized concept must be translated to a standard number of fish per unit volume in order for it to be used as an effec tive management tool.

At present, mostly from a lack of comprehensive data of the nature described above, no universal standards exist. The most practical approach then is to apply a so-called "sliding scale" of stocking densities that changes as fish age (Matthews et al. 2002). These authors suggest that densities of 40–50 fish per liter are appropriate during the early larval stage, but should be gradually reduced to 5 fish per liter for juveniles and adults. This is quite conservative, relative to standard stocking densities of other cultured fish species, and therefore is an acceptable guideline that may be used to inform management until more specific studies on the subject are published.

genetic management

Laboratory stocks of zebrafish are typically maintained as relatively small, closed populations that are subject to continuous losses of genetic diversity resulting from founder effects, genetic drift, and population bottlenecking (Stohler et al. 2004). In such situations, the risk of inbreeding depression is increased because mating between close relatives is much more likely. Populations that suffer from inbreeding depression are typically characterized by poor reproductive performance, reduced growth and survival of offspring, and increased susceptibility to disease outbreaks. Therefore, it is crucial to limit inbreeding to the greatest extent possible in order to preserve the long-term health and productivity of laboratory populations.

There are three important ways in which genetic diversity in laboratory stocks can be maximized: (1) the maintenance of large effective population sizes, (2) strict avoidance of breeding between siblings and close relatives during propagation events, and (3) periodic importation and crossing of individuals from an outside source population into existing stocks **(Figure 31)**.

Effective population size is a term that refers to the number of individuals in a population that actually contribute offspring to the next generation. This number can be maximized during stock propagation by selecting offspring derived from either many pair crosses or group mating events involving large numbers of individuals. Using pair crosses for this purpose is preferable, because it allows for greater control of parentage and the avoidance of breeding closely related animals. Group matings are less than ideal because dominant individuals may monopolize fertilizations and disproportionately contribute to new generations, thereby actually reducing effective population size.

When new stocks are propagated, great care should be taken to ensure that the animals used to generate new progeny are not siblings or closely related to one another. This is only possible if the parentage of the broodstock is known and controlled, at least up until the point that the animals are bred to make a new generation. To achieve this, broodstock should be reared and housed discretely according to pedigree to sexual maturity. For the propagation event, they can then be set up in crossing events with non-related individuals to generate new stocks.

Over time, the genetic diversity of closed populations may decline to the point where inbreeding depression problems occur, even if matings

1.

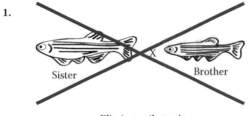

Eliminate sib matings

2.

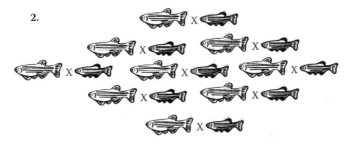

Maximize effective population size

3.

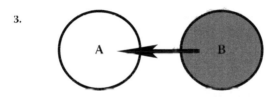

Periodic importation of individuals from outside populations

Fig. 31 Combined strategy for maintaining genetic diversity of laboratory populations. The genetic diversity of laboratory populations of zebrafish may be maximized by a combined strategy that includes (1) strict avoidance of mating events between siblings, (2) maximizing effective population sizes, and (3) periodic importation of individuals from outside populations.

between siblings are scrupulously avoided and effective population sizes are high. It is possible to offset the resultant reductions in diversity by importing fish from outside populations of the same strain so that they can be crossed with existing stocks. These outbreeding events will serve to infuse existing populations with new levels of genetic diversity, as fish from disparate populations carry different alleles that will be contributed to the gene pool. For the purposes of maintaining diversity over the long term, stocks can be outbred to fish from outside populations once every several generations, or as soon as effects of inbreeding depression become apparent.

The aforementioned strategies are most appropriate for managing populations of wild-type stains. Mutant and transgenic strains can be maintained by either incrossing fish that carry the mutation or transgene or by outcrossing them to wild-type fish (Westerfield 2007). Although incrossing will yield higher numbers of carriers in the next generation, outcrossing is preferred because it maximizes genetic diversity. In either case, efforts should be made to use as many carriers as possible when making new generations to guard against genetic bottlenecking. Finally, care should always be taken when outcrossing mutant or transgenic stocks to wild-type strains different from the one that they were originally generated on to ensure that the transgene or mutant phenotype is faithfully transmitted in the new progeny, as the expression of the phenotype may be altered in the new background (Sanders and Whitlock 2003).

identification and record keeping

Identification

An essential element of zebrafish facility management is the proper identification of fish. While no practical method exists for the tagging of individual animals, the labeling of specific groups of fish is a relatively straightforward organizational endeavor that has important implications for stock management, compliance with animal care and usage regulations, and billing.

Tank labels

The most basic fish identification tool is a tank label, which is a smaller and often less formal version of the cage card typically used in traditional animal facilities. The key physical requirements of a tank label are that it be resistant to water and that it fit on the front of even the smallest of tanks. The best type of label to use for this purpose is one that has been generated by a professional label maker or printer on self-laminating labeling tape because the print does not fade even after exposure to moisture, and is universally legible (as opposed to someone's handwriting) **(Figure 32)**. It is also possible to use water-resistant labeling tape that can be written on

Fig. 32 Printed tank labels on zebrafish isolation tanks. Labels generated by label printers are preferable to handwritten labels because they are easier to read and do not fade over time.

with a permanent marker, although extra care must be taken to ensure that the information on labels of this type remains readable over time.

There are several categories of information that should be on a tank label. It is important not to include too much or too little; the goal is to be as clear and consistent as possible. The required fields on a tank label should be

- *Date of birth:* The date of birth, preferably given in universal fashion (dd-mm-yyyy) should always be on the label, as this will allow for ready assessment of the age of the animals during their existence in a facility.

- *Name of the responsible investigator:* The name of the specific investigator or laboratory group should always be on the label so that the party responsible for the animals is evident. The animal protocol number associated with the fish should also be included. It may also be acceptable to use different colors of labels to indicate ownership.

- *Strain name:* The name of the strain, including any particular information regarding the particular pedigree of the animals, should be clearly displayed on the label.

- *Stock number:* If a stock-numbering system is used to track fish (see below), the number should be displayed on the label.

Record Keeping

Good record keeping is imperative for efficient zebrafish facility management. A wide variety of information should be regularly collected, recorded, and kept on hand in logbooks that can be easily reviewed if the need arises. Data should always be collected consistently and recorded in such a way that the information is readily apparent to anyone viewing it. Notable items to keep records of include the following:

- *Water quality:* Key water quality and environmental parameters, including pH, salinity, temperature, and nitrogenous wastes, should be recorded in a logbook **(Appendix—sample water quality checksheet)**. This also includes activities related to the maintenance of water quality, such as the filling of salt and bicarbonate dosing reservoirs and changing of filters. Collection of this type of information allows for the ready analysis of water quality trends and rapid troubleshooting when values move outside target ranges.

- *Equipment maintenance and repair:* The routine maintenance of all system equipment should be recorded in logbooks. This includes activities such as the replacement of UV bulbs and quartz sleeves, the calibration of monitoring system probes, and filter changes. Records should also be kept of when equipment components fail or malfunction, as well as when they have been repaired or replaced.

- *Feeding:* Records of all activities related to feeding and feed management should be kept to ensure that stocks are being fed according to established standard operating procedures and daily schedules. It is also advisable to keep track of when feed items are received from suppliers so that they can be used before they expire.

- *Fish tracking:* A simple stock-numbering system can be employed to help track fish numbers, usage, and movement in a facility. This typically involves the assignment of a unique, sequential stock number that identifies the fish and stays with them during their lifetime in a facility. The stock number can be easily integrated into a larger recordkeeping program that tracks movement of the fish from place to place in a facility, as well as health, breeding status, and mortality. This process can also be readily automated by using a standard bar code reading system.

- *Movement:* The movement of fish within and outside of a facility should be recorded each time it occurs. Ideally, the location of each family or distinct group of fish in a facility should always be known. This is most easily utilized in conjunction with a stock numbering system.

- *Retirement:* The retirement or euthanasia of family or distinct groups of fish should be recorded when it occurs. This is most easily utilized in conjunction with a stock-numbering system.

- *Genetic management:* Records on the pedigree of families or distinct groups of fish should be maintained to facilitate the genetic management of stocks. Collected information of this nature should include parentage and mating schemes for each new generation of fish.

- *Breeding:* A record of the reproductive performance of all breeding fish should be kept. This may include the frequency of spawning events, success or failure of crosses, and fecundity and fertility rates. Such information can be utilized to help ensure the proper cycling of breeding stocks, and can aid in the monitoring of reproductive trends in fish populations.

- *Mortality records:* All mortalities that occur in a facility should be recorded. The information that should be captured in such a record includes the date and time of mortality (or collection of carcasses), strain, location, and number of dead fish collected. Pertinent observations concerning the dead animal(s) as well as any live fish that remain in tanks where mortalities have occurred should also be noted and logged. In the event of disease outbreaks, these data can aid the identification, diagnosis, and treatment of the particular disease agent at work.

- *Import/export:* All fish that are imported into and exported from a particular facility should be recorded. For imports, the collected information should include the source of the fish, date of receipt, quarantining methods employed, strain, age, and number of fish received. Notes on the condition of the fish upon arrival are also useful. For exports, it is important to note the date of shipment, as well as the number, age, and strain of fish being sent out.

transportation

As the zebrafish model system continues to grow in popularity and scope of usage, so too does the demand for wild-type, mutant, and transgenic strains of fish for use in laboratories across the globe. Consequently, the practice of shipping live zebrafish between institutions—both within and outside domestic borders—has become an increasingly common activity.

This presents a number of challenges. Transport is physiologically stressful for fish, especially over long distances. There are also complex and evolving regulatory requirements governing the movement of live animals across international borders that must be navigated. Finally, the exchange of fish between laboratories always presents a risk for the spread of pathogens, whether the facilities are next door to each other or are on different continents.

To deal with these issues effectively, veterinarians, facility managers, and others charged with the oversight of zebrafish research programs need to be well versed in the various principles of fish shipping and transport.

Shipping

General physical requirements

The principal challenge of transporting live fish is the maintenance of suitable water quality in shipping containers during transport. The most crucial parameters are ammonia and dissolved oxygen. As a result of normal metabolism, fish consume oxygen and produce ammonia and carbon dioxide. While this is generally not an issue in a well-managed recirculating system, the levels of dissolved oxygen and ammonia can be highly problematic in the context of a closed, static entity such as a shipping container. With no input of fresh water or means to process and eliminate wastes, concentrations of dissolved oxygen will decrease and ammonia will increase over time. The rates of both of these increase with fish densities. The key to the proper packaging of fish is to minimize these changes during transport, such that dissolved oxygen levels do not fall below critical values and that ammonia does not exceed toxic levels.

The most straightforward way to deal with this problem is to reduce densities of fish in shipping containers. Fewer animals consume less oxygen and produce less ammonia, so when densities are low, the margin for error is greater. Therefore, packing zebrafish at low densities is always the recommended option, especially since there is usually little need to ship large numbers of animals for research purposes.

Other simple measures can be employed to ensure maintenance of favorable water quality during transport. Suitable levels of dissolved oxygen can be maintained by packing animals in containers with a volume of air that is 3 to 4 times the volume of transport water (Lim et al. 2003). In ornamental fish production settings, pure oxygen is often added to the bag, but in typical zebrafish situations (lower densities), the addition of simple air is acceptable. Oxygen consumption and ammonia production rates can be reduced during transport by fasting fish for 24 to 36 hours prior to packaging. Ammonia levels can further be stabilized by the direct addition of ammonia-binding products, such as Amquel® (Kordon, LLC), to the transport water, although this is not necessary when shipping fish at low densities.

It is also very important to maintain stable temperatures in transport water during shipment. If the water becomes too warm, it will hold less dissolved oxygen and the toxicity of ammonia is increased. If the temperature gets too low, the fish may become hypothermic. The transport water can be buffered against drastic temperature changes in several ways. The use of thick-walled Styrofoam shipping boxes helps insulate the fish from changes in outside temperatures, as well as protect them from the physical disturbance of jostling during transport. Extra bags of water can also be added to the box to help moderate temperatures in the shipping container. When shipping to cold-weather climates or during the winter, small heat packs can be added to the box to maintain adequate temperatures.

Sample packaging method – adults

1. Prior to being shipped, the fish should be checked for any visible signs of disease or distress. If the animals are healthy, they should be placed in a special tank on system, and feed should be withheld for at least 24 hours prior to shipment.

2. On the day of shipment, a plastic shipping bag (these can be purchased from any aquaculture supplier) should be filled with fresh, clean water to approximately 30–40% of the total capacity **(Figure 33)**.

3. The fish should then be removed from their tank and added to the bag at a density of no greater than 5 fish per liter **(Figure 34)**. The bag should then filled with either air or pure oxygen (if available) and then sealed by twisting the top shut, folding the "twist" over on itself, and then closing it with at

Fig. 33 Shipping bag prior to addition of fish. Plastic shipping bags should be filled with fresh water to approximately 30% of their full capacity prior to the addition of fish.

least two thick rubber bands **(Figure 35; Figure 36)**. A label identifying the fish should then be added to the outside of the bag **(Figure 37)**.

4. The bag should then be placed, upside down, inside of another empty bag so that the seams of the inside bag containing the fish align with the seams of the outside bag **(Figure 38)**. The outside bag should then be sealed in the same manner that the first bag was. Care should be taken to ensure that no air is trapped inside the second bag **(Figure 39)**.

5. The bag should be transferred into a standard shipping container that consists of two boxes: a thick-walled Styrofoam inner box that fits snugly inside a slightly larger cardboard box **(Figure 40)**. The bag should be placed inside the inner box, which can be filled with bubble wrap and/or packing peanuts to ensure that the bag does not move or shift around during transport **(Figure 41)**. If heating packs are used, they should be placed either in the bottom of the box beneath a layer of bubble wrap/packing peanuts, or affixed to the underside of the lid of the box, again insulated by a layer of packing material. These heating elements should NOT be in direct contact with the bag.

Fig. 34 Shipping bag with fish, prior to sealing. Fish should be added to the bag at densities no greater than 5 fish per liter.

6. The lid should be placed on top of the inner box and sealed tightly with packing tape **(Figure 42)**.

7. The sealed inner box should then be placed inside the outer box. The outer box should then be closed and sealed tightly with packing tape. The shipping label and associated documents (if required) can then be affixed to the outside of the box. The container is now ready for shipment **(Figure 43)**.

Fig. 35 Sealing method of fish shipping bag (a). The bag should be filled with air, and then tightly twisted shut.

Sample packaging method – embryos

1. Fertilized embryos should be collected from spawning crosses, cleaned, and stored overnight at 28°C.

2. At 24 hours postfertilization, non-viable eggs should be removed, and the remaining viable eggs should be bleached according to standard protocols (Nusslein-Volhard and Dahm 2002; Westerfield 2007).

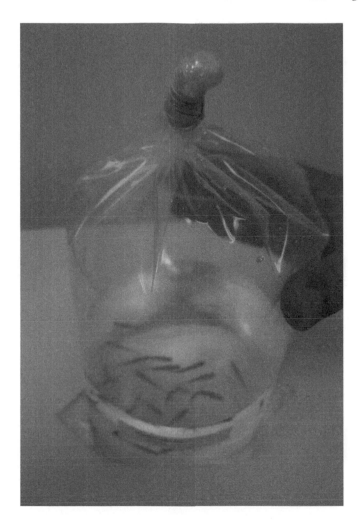

Fig. 36 Sealing method of fish shipping bag (b). The twist should be folded onto itself, and sealed with two rubber bands.

3. Embryos should be transferred to sterile 50-mL conical tubes or 250-mL tissue culture flasks that are filled to 80% of the total available volume with autoclaved fish water or embryo medium **(Figure 44)**. Densities should be no greater than 2 embryos/mL. In some instances, a low dose of methylene blue can be added to the water to help reduce fungal growth in the water during transport (Westerfield 2007).

4. The flasks or conical tubes should be closed and sealed with parafilm to prevent leaking **(Figure 45)**.

Fig. 37 Fish shipping bag, sealed and labeled.

5. The flasks or tubes should then be placed inside the inner box of the shipping container. Packing peanuts and/or bubble wrap can be used to insulate the contents and prevent them from moving during transport **(Figure 46)**.

6. The lid should be placed on top of the inner box and sealed tightly with packing tape.

7. The sealed inner box should then be placed inside the outer box. The outer box should then be closed and sealed tightly with packing tape. The shipping label and associated

Fig. 38 Fish shipping bag inside a second bag. The sealed, labeled bag should be placed upside down inside a second bag, with seams aligned.

documents (if required) can then be affixed to the outside of the box. The container is now ready for shipment.

Receiving

General physical requirements

There are a number of factors to consider when receiving fish, particularly adults. When fish first arrive at their destination, it is critical that they be inspected immediately and then transferred from shipping containers into clean, fresh water. Expedience during this process is particularly important for fish that have been densely packed or have been in transit for over 24 hours.

Fig. 39 Fishing shipping bag, inside sealed outer bag. The outer bag should be sealed in the same manner as the inner bag, but with excess air removed.

While it is crucial to work quickly, it is equally imperative to be cognizant of the nature of the chemical changes that take place in shipping containers during transport, as this defines how the fish should be unpacked and transferred. During normal metabolic function, fish absorb oxygen and excrete carbon dioxide and ammonia. In the static, closed environment of a shipping bag, the concentrations of both metabolites accumulate in the water. The toxicity of total ammonia nitrogen (TAN) in the water is dependent upon its pH and temperature, as the fraction that is toxic—NH_3—increases with pH and temperature. During transport, the carbon dioxide released by

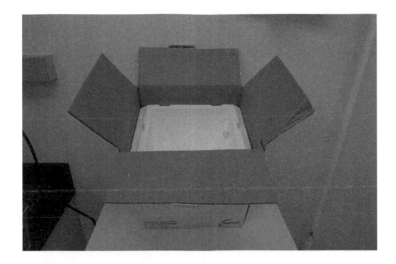

Fig. 40 Standard fish-shipping container. A standard fish-shipping container consists of a thick walled Styrofoam box inside an outer cardboard box.

Fig. 41 Fish-shipping bag placed inside shipping container. The fish-shipping bag should be placed inside the shipping container, and insulated with packing peanuts.

Fig. 42 Inner Styrofoam box, sealed with packing tape.

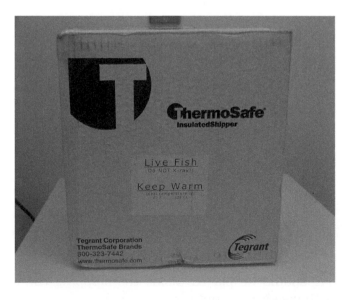

Fig. 43 Fish-shipping container, sealed and ready for shipment.

Fig. 44 Tissue culture flask, filled with water. Tissue culture flasks are a preferred vessel for shipping zebrafish embryos.

Fig. 45 Tissue culture flask, sealed with parafilm. Flasks should be sealed tightly, and then wrapped with parafilm (black arrow) to help prevent leakage.

the fish reacts with water to form a weak acid, which lowers its pH. This serves to keep the toxic fraction of TAN that is present in the water at lower, less problematic levels.

This all changes once the shipping bag is opened and exposed to the environment. The CO_2 that has accumulated in the bag is immediately released into the atmosphere, which results in a quick rise in the pH of water. This increase in pH in turn causes the toxic fraction

Fig. 46 Tissue culture flasks inside standard shipping container. Flasks should be insulated with bubble wrap.

of total ammonia nitrogen to spike, which can severely stress or kill fish, especially those that have just been subjected to the rigors of transport.

To avoid this scenario, shipping bags should not be opened until the fish are ready to be transferred. Once the bag is opened, the fish should immediately be poured out into a net and transferred to new water. This step is most critical when shipping densities and temperatures are high, but should be a necessary part of any receiving protocol.

It is also important to acclimate the fish for temperature upon arrival. Fish should not be removed from their shipping containers until the temperature of transport water is the same as the receiving water to avoid complications arising from thermal shock.

Biosecurity concerns

Fish imported from outside sources—whether from research laboratories or commercial vendors—may carry pathogens that pose a serious risk to the health of populations in an existing facility. Precautions should be taken with all incoming animals to ensure that biosecurity is not compromised. The simplest way to achieve this is to limit the entry of any animal from the outside to embryos that have been surface disinfected with bleach. Non-bleached embryos and all other life stages of fish should be placed in and kept in a separate quarantine facility for the rest of their existence. This way, most foreign pathogens that the fish may harbor will be contained, as long as standard quarantine procedures are strictly followed. While in quarantine, these fish can be used for experiments and/or to produce embryos that can be bleached before transfer into a main facility.

There are a number of published egg-bleaching protocols that are widely used in the zebrafish research community (Nusslein-Volhard and Dahm 2002; Westerfield 2007). Typically, embryos are bleached with chlorine levels ranging from 25 to 50 ppm for 2 to 10 minutes to prevent transmission of pathogens. This practice is excellent for removing most external surface organisms. However, certain microsporidia spores, such as *Pseudoloma neurophilia*, are highly resistant to chlorine, and the 25- to 50-ppm chlorine protocol used in most zebrafish research facilities will not prevent transmission of *P. neurophilia* (Ferguson et al. 2007).

Sample receiving method – adults

1. Open the outside box of the shipping container immediately upon receipt. Remove bag from inner container.

2. Float the bag in the receiving water, or in water that is the same temperature as receiving water, for at least 5 minutes per degree of temperature difference or until the temperature of the bag is within 2 degrees of the receiving water (Cole et al. 1999).

3. Remove the bag from the water, cut a hole in it, and then, over some kind of container or bucket, pour the fish out of the bag into a net. Transfer them into the receiving water. Discard the transport water.

4. Observe the fish. Note any overt sign of stress, illness, or trauma. Take action (remove and treat, or euthanize), if necessary.

5. Allow the fish to acclimate to their new environment for at least 24 hours before feeding.

Sample receiving method – embryos

1. Open the outside box of the shipping container immediately upon receipt. Remove tubes or flasks from the inner container.

2. Open the tubes or flasks, and pour water and embryos/larvae into petri dishes.

3. Transfer viable embryos/larvae into fresh fish water or embryo medium. Discard transport water and debris.

4. Incubate at 25–30°C until the fish are placed in a nursery system or used for experiments.

Regulatory Requirements

At the time of this writing, the transfer of zebrafish between institutions and laboratories domestically is generally very loosely regulated. Typically, the only documentation that must be completed for such shipments is the specific forms requested by the particular institutions participating in the transfer, if any are required at all. Still, it is highly recommended that shippers consult local governmental regulations to ensure compliance before transfers are expedited.

The practice of shipping fish across international borders, however, often requires that considerable documentation be completed before the transfer can actually take place. The nature of this paperwork varies considerably, depending upon the requirements of the particular parent countries of the institutions involved in the transfer.

At minimum, the shipping of live zebrafish embryos or adults across international borders requires that a customs declaration be completed by the shipping agent and included with the shipment. This general document, which is reviewed by customs officials in the country of receipt, discloses the contents of the container and the nature of the transfer being undertaken. Some countries may also require that the recipient apply and receive approval for imports prior to their clearing customs. In some instances, health certificates, signed off on by a veterinarian in the country of origin, may also be necessary for the importation to be completed. In all instances, it is advisable that the involved parties employ a licensed courier service to help them successfully complete the import/export documents required for the shipment.

life support

facility design

In terms of its intended functionality, a zebrafish facility has much in common with a typical mammalian laboratory animal facility. The facility must be equipped in such a way that it controls and maintains environmental conditions, limits the spread of pathogens, and facilitates care as well as research workflow. Beyond these basic features, however, there are few similarities. Many of the differences between fish and mammal facilities have to do with water—the medium that sustains life in any aquatic facility. Fish facilities are dynamic entities that revolve around water, which both delivers (oxygen, minerals, food, etc.) and removes (wastes, pheromones, etc.) materials to and from the fish. Consequently, and above all, fish facilities need to be outfitted to handle water in a wide variety of ways.

Water Handling and Resistance

Zebrafish facilities are inherently "wet." Fish housing systems and tanks hold large volumes of circulating water, with associated pumps moving water through plumbing from water source to tanks and back again. Fish facilities are also high traffic areas, with many research and animal care personnel working in rooms throughout the day. Such work requires the routine movement of fish on and off systems, and from tank to tank. Finally, the routine maintenance of the mechanical components of fish systems also involves hands-on and often "wet" work. Consequently, zebrafish facilities are seldom dry.

Given this reality, all materials used in the construction of fish facilities must be designed for use in moist environments, and exposed surfaces should be resistant to corrosion and rust. Equipment that is by design or functionality less suitable for constant exposure to water (such as electrical outlets) should be contained in sealed, watertight compartments.

The tendency for zebrafish housing and procedure rooms to retain standing water presents not only an occupational health issue for the people working in them, but also a health risk to the fish populations that they support. Water containing pathogenic organisms may expose uninfected animals in a variety of ways, and the risk of such an event is further heightened by standing, stagnant water in which bacteria, protozoa, or other organisms may proliferate. To offset these issues, surfaces (floors, counters, bench tops, etc.) should be impervious to moisture and easy to clean. All rooms should have at least one floor drain **(Figure 47)**, and floors should be pitched toward drains to eliminate puddles resulting from routine work in the room. Design should also facilitate cleaning; difficult to access areas in and under racks could also pose hazards in terms of harboring reservoirs of pathogens or other materials that could be harmful for fish or personnel.

Fig. 47 Floor drain.

Water Production

Most, if not all, fish facilities must pre-treat the water before running it into fish housing systems. This can involve reserve osmosis filtration and/or storage of this water for use in the fish facility. It is ideal to place this water production equipment in dedicated space, isolated from housing areas. This facilitates maintenance and reduces noise and vibration associated with filtration and distribution.

Infrastructure

There are a number of items to consider in terms of the overall infrastructure of a fish facility. As fish systems (tanks and water) are quite heavy, the dead load capacities of floors should be designed to easily support the weight of racks and water. It is advisable during the planning phase to ensure that flooring exceeds projected weight capacities so that future expansions can be supported if those needs arise. It is also recommended that racks be secured with earthquake restraints to protect fish and personnel (Bailey 2008).

Given the considerable volumes of water moving through a typical zebrafish facility at all times, it is crucial that rooms be designed to contain the floods that will inevitably occur from time to time through human error, equipment failure, or some other disruption. All rooms with fish or system-related mechanical equipment located in them should have floor drains. An additional level of protection can be afforded by installing flood sensors that are designed to send alarms to building operations or facility staff when water levels on the floor exceed some threshold level **(Figure 48)**. It is also advisable to keep a wet-dry vacuum on site to help with cleanup when the need arises.

Design should also address infrastructural maintenance concerns. Plumbing, electrical controls, and other important mechanical components of systems should be easily accessible to facilitate maintenance and repairs. Equipment should also be located in zones that are located out of the way of normal workflow. Facilities designed to have separate and contained mechanical space are preferable for this reason, as well as to minimize noise, heat, and vibration in housing and work areas.

Electricity

Zebrafish facilities are energy intensive. Pumps, lights, ultraviolet sterilizers, filters, and other mechanical components all require

Fig. 48 Flood sensor. Flood sensors located at various points in fish rooms are an extra layer of protection against flooding.

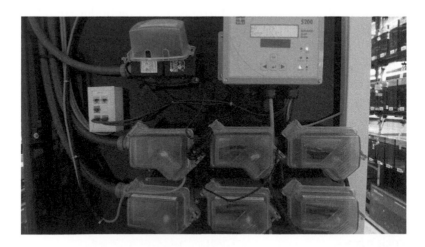

Fig. 49 Covered electrical outlets. Electrical outlets should be designed to operate in moist operating environments and should be sited away from high water flow areas (receptacles in the photo are placed at end of aisles, mounted on plastic shields and out of direct contact with tanks).

substantial and uninterrupted electrical power. All electrical equipment must be installed according to local codes so that they can be safely operated in moist environments (Canadian Council on Animal Care 2005). Circuits should be grounded, and wires and electrical outlets should be contained in water-resistant containers or receptacles and located away from high workflow areas and out of direct splash zones **(Figure 49)**. Submersible electrical equipment (such as heaters) should be properly designed to operate in underwater conditions. Importantly, all critical equipment (pumps, ultra-violet sterilizers) should be backed up by emergency power sources that remain functional in the event of power failure.

Environmental Controls

Fish housing rooms should be designed so that environmental conditions can be tightly controlled and monitored. Room temperature must be regulated within a reasonably narrow range, with minimum- and maximum-temperature-alarming capabilities to ensure that swings of greater than ±3–5°C do not occur. Air-handling systems should ensure maximal ventilation in housing areas, minimizing humidity and aerosol transfer in rooms. Lighting should also be regulated; photoperiod is critical for fish in terms of regulating growth and reproduction, as well as for the maintenance of live feed cultures such as *Artemia*. A light–dark cycle of 14:10 is most commonly employed. More sophisticated facility design incorporates dimmers in lighting controls so that lights come on and go off gradually, a condition more reflective of natural conditions (dawn and dusk).

Light intensity is also important. Although the optimal intensity for zebrafish has not yet been defined, the general recommendation is for it to be between 5 and 30 foot candles, or 54–354 lux, at the water surface (Matthews et al. 2002).

Alarming

As the fish facility contains critically important (and expensive) resources, design should include the capacity to protect animals and equipment and maintain conditions independent of outside events. Room parameters should be automatically monitored on a 24/7 basis with minimum/maximum threshold alarms for temperature, light, and water levels and physical parameters. All critical equipment should be or have the ability to be rapidly connected to emergency backup power supplies that protect against loss of function during local power outages.

Design should also incorporate redundancy in equipment (pumps, UV-sterilizers water storage units, etc.) whenever possible so that critical components can be readily switched out in event of failure.

Access

Access to fish facilities should be restricted to fish care and research staff, and to maintenance personnel specifically trained to work in fish facilities. Limiting foot traffic to essential personnel is good practice because it minimizes overall disturbance in the rooms, reduces pathogen transmission, and protects sensitive equipment. Facilities can be specifically designed to adhere to this strategy, with controlled access to animal holding facilities and mechanical space. If electronic card readers **(Figure 50)** are utilized, more sophisticated approaches can be employed, including limiting access to specific time periods for subsets of affiliated personnel, and entry counts and limits that may aid with quarantining procedures (once someone has entered quarantine space, that person may not reenter the main holding space within some defined period).

Facilities should have only one main access point. This area or zone should be a vestibule or foyer that may include room for footbaths, hand cleaning, and the storage and application of dedicated footwear or clothing. If building codes require more than one egress point for the room, it can be managed such that it is used in only the event of an emergency.

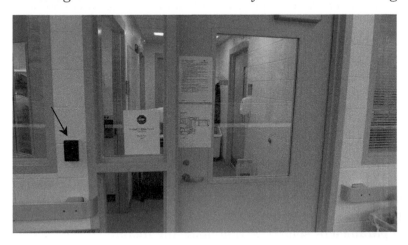

Fig. 50 Card access reader. Card swipe access readers (black arrow) at facility entry/exit points can limit access to fish rooms to affiliated personnel only.

Work Zones

Zebrafish facilities are user-intensive spaces. At any given time, a number of personnel may be working in rooms on research, fish care, or system maintenance. This setting, if not properly controlled, may lead to diverse operational problems, including congestion and reduced workflow, as well as enhanced transmission of pathogens between separated groups of fish. Consequently, it is preferable for fish housing rooms to have specific work zones to ensure that such diverse activities can all be supported simultaneously without compromising the health of the fish. To achieve this, rooms should be designed to have simple, discrete work zones. Designated spaces for breeding cross setup and takedown, tank cleaning and sterilization, and general procedures should be established. Pumps and filters and other water-treatment-related equipment should be located away from housing and procedural zones to help reduce noise and vibration—which can be disruptive to fish health and experiments (see Chapter 6)—as well as to ensure that regular maintenance and repairs can be performed without interrupting normal workflow. Separate rooms should be provided for storage, feed preparation, and scientific procedures (microscopy, microinjection, imaging, etc.) **(Figure 51)**.

Storage

There are considerable equipment needs in any fish facility. Rack and system components, tanks, breeding chambers, feed-culturing materials, plumbing supplies, and numerous consumables (salts, chemicals, feed additives, gloves, basic labware, etc.) are all required and need to be on hand for normal operation in any facility. Facilities should be designed such that all of these items can be stored in space that is physically separated from fish-holding and work areas.

Rack Configuration

The configuration of fish system racks is an important consideration in facility design. The overriding principle of fish facility design with respect to rack configuration has historically been to fit as many tanks as possible within a given space footprint, but this is a narrow and oversimplified approach that limits the possibilities of future use. Progressive design should consider a number of concepts in configuring rack layout in rooms. Maximizing workflow should be a priority; wide aisles, discrete work zones, and locating activities such

Fig. 51 Procedure room. Space for procedures should be separate from fish housing areas.

as procedures, feed preparation, equipment maintenance, and storage away from fish housing are all ways to enhance the efficiency of space. Flexibility is another; modular racks that can allow for various configurations of tanks of varying sizes, depending upon experimental situations, are also highly advantageous in this regard. It is also advisable to plumb racks or groups of racks so that they may be taken on and off line or isolated from other racks as needs arise (Bailey 2007).

Quarantine

A crucial component of biosecurity in zebrafish facilities is quarantine space to hold fish imported from outside sources. Fish facilities should all be designed to incorporate discrete quarantine rooms that are physically separate from main holding rooms. Quarantine rooms should be completely isolated from main holding rooms; they should at minimum have a discrete water supply, equipment, sterilization/ cleaning areas, and procedure space. Access to this space should be limited and preferably controlled such that only essential personnel enter the room. Outfitting rooms with some mechanism for timed entry may provide an additional layer of security; once personnel enter quarantine space, they may not access main holding rooms for some predetermined period. Simple standard operating procedures can also be used to direct workflow in this fashion. More details on the benefits and operation of quarantine space are provided in Chapter 5.

Air

Compressed air, preferably sterilized in some way before entry into fish rooms, should be made available throughout fish-holding and wet-procedure rooms. Air is needed for a variety of applications in fish facilities. Air may be used to off-gas chlorine that may be present in municipal water sources, to circulate water in static procedure tanks, and to aerate live feed cultures. It is of course possible to use stand-alone pumps or compressors for these purposes, but outfitting rooms with building air maximizes space efficiency and provides future options for users.

Sinks

Given the nature of the work done in any zebrafish facility, the inclusion of sufficient numbers of sinks that are both deep enough and wide enough to contain large volumes of water is of crucial importance. Ideally, there should be numerous sinks located throughout fish-holding and affiliated rooms, each with ample adjacent counter space that allows for staging and breakdown of tanks and equipment. Sinks should have deep basins and be equipped with high-pressure spray hoses to facilitate cleaning of tanks and nets **(Figure 52)**. Additionally, faucets should be plumbed to source water so that "fish" water can be accessed in these spaces for the purposes of filling tanks with the appropriate water.

Cleaning and Sterilization

The design of fish facilities must incorporate the means to clean and sterilize large volumes of equipment in support of daily operations. Commercial undercounter glass-washers, which are the most commonly utilized cleaning machines in this setting, can easily be included in the sink areas of fish holding and procedure rooms, preferably in a physical arrangement that supports logical workflow of dirty equipment directly into washers after cleaning or spawning breakdowns **(Figure 53)**. If larger-scale cleaning equipment (e.g., cage or tunnel washers) is used, it should be located within the room/ suite or immediately adjacent to it so that equipment going to and from the area travels minimal distances. Decreasing travel distances

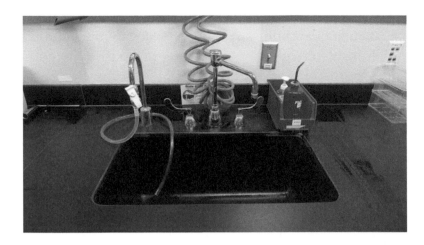

Fig. 52 Dirt sink. Fish room sinks should be deep basined for wet work and cleaning.

Fig. 53 Undercounter glass washer. Undercounter glass washers are standard cleaning/sterilization equipment for fish facilities.

is important because lengthening distances that workers may have to travel to and from cleaning and sterilizing equipment may present problems for pathogen control, and could pose occupational hazards through spills and leaks in traffic areas.

Redundancy

The equipment in fish rooms operates continuously, and is thus prone to periodic breakdown and failure. Since such interruptions pose a serious threat to the fish and research being conducted in the facility, it is important to plan for such events by incorporating redundancy into design. Such redundancy allows for quick replacement and resumption of vital support system function in the event of failure. This extra level of coverage should include backup power sources to run equipment and life support, backup pumps for systems and water production applications, and the ability to monitor and regulate water temperature and chemistry via more than one pathway (for temperature HVAC or electric heaters).

housing systems

While it is of course possible to maintain zebrafish in traditional glass aquaria, their greatly expanded use as a model organism has necessitated the development of something more sophisticated. Consequently, a number of companies have sprung up to meet the growing demand for research-grade fish housing systems. These vendors have melded concepts from traditional aquaculture, rodent housing, and research genetics to create sophisticated life support systems for the housing of zebrafish in laboratory settings. These systems are an elegant work of aquacultural engineering: a mix of tanks, racks, and filtration driven by complex biological, chemical, and mechanical processes.

At their most elemental level, zebrafish housing systems comprise three basic building blocks that serve to provide clean water for fish: **tanks**, **racks**, and **water treatment**. Racks provide the structural support for the tanks that house the fish as well as for the plumbing that delivers clean water to and receives wastewater from the tanks. Pumps drive the water through the system. In most cases, wastewater is filtered and cleaned within the system before being returned to tanks. A breakdown and description of these components is provided below.

Tanks

Although tank design varies with vendor, all tanks generally are constructed to function in the same manner **(Figure 54)**. Clean water
is delivered to tanks through a hole in the tank lid by means of a
tube or specialized valve that allows for modulation of water flow.
The direction of flow into the tank circulates the water inside, and
flushes solids (uneaten feed, feces, etc.) out of the tank through an
overflow drain, which may be located in the back or front of the tank,
depending on system vendor. All tanks must contain some provision to prevent fish from swimming or being washed out the drain.
Some tank designs achieve this through the use of a baffle, which is
a plastic insert that fits snugly inside the back wall of the tank and
in front of the drain and serves to prevent fish from escaping while
facilitating waste removal. A slight gap at the bottom of the baffle
allows wastewater and solids to leave the tank but is too small for
adult fish to fit through. For small sub-adult, juvenile, or larval fish
that could swim under the baffle and escape, specialized baffles that
sit flush with the bottom of the tank are employed. These so-called
"baby baffles" contain screens in the center of the piece that facilitate
water and waste movement and prevent fish loss to the drain. In a
fewer number of tank types, there is no baffle; the drain or tank hole
may be simply covered directly with mesh screen that prevents fish
from escaping. The size of the holes in the mesh is changed according

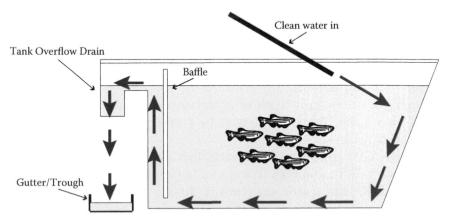

Fig. 54 Typical tank design. Water is delivered into the tank via inlet
tubing or hose. Direction of water flow in the tank (blue arrows) is
into the front of the tank, along the bottom, up behind the baffle, out
of the overflow drain, and into the return gutter/trough.

TABLE 4: TANK DESIGNS AND USES

Tank Type	Typical Volume	Usage	Typical Capacity
Isolation	0.75–1.75 liters	Isolation	1–2 adult fish
Nursery	0.75–3 liters	Larval rearing	Up to 100 larval fish
Small adult holding	2.75–4 liters	Juvenile grow-out, adult housing	Up to 20 adult fish
Large adult holding	8–12 liters	Juvenile grow-out, adult housing	Up to 100 adult fish

to the size of the fish housed in the tanks; larval tanks will require a much smaller gauge size than adults. Both screens and baffles require periodic cleaning, as solids will eventually accumulate and clog drains.

Most tanks are made from clear polycarbonate, although they may also be composed of glass, acrylic, or polysulfone. The fact that polycarbonate is the most commonly used component for zebrafish housing tanks is an issue of potential concern; it has been shown recently that a primary component of polycarbonate, bisphenol A (BPA), is harmful to vertebrate animals, including zebrafish, when present in high enough concentrations (Duan et al. 2008). It remains to be seen if circulating levels of BPA leaching from system tanks are high enough to impact research in a negative way. In realization of this potential issue, at least one system vendor has recently begun to offer BPA-free tanks to customers. All tanks, except for systems with fixed glass runways where tanks may not be readily taken on or off system, may be removed from racks for the purposes of cleaning and sterilization.

Zebrafish housing tanks come in a variety of sizes and shapes, although they may be generally grouped in three or four size classes or types, based on intended function **(Table 4)**. Realistically though, the size of the tank used will be dictated by personal preference and experimental application. Tanks are outfitted with tight-fitting lids, which may be colored blue or green to help prevent algal and/or cyanobacterial growth.

Racks

Racks provide structural support for tanks and water delivery, removal, storage, and treatment. The prototypical rack "unit" is a stainless steel structure divided into a number of "shelves" upon which rows of tanks are aligned. The exact configuration of the rack will vary to a certain extent depending on vendor and available space, but as an example, a typical space footprint for a 6-shelf unit would be 60"L × 14"W × 90"H. Each shelf in a rack is outfitted with supply piping and a gutter. The supply piping typically runs along the top of the shelf, and delivers clean water to the tanks via flexible plastic tubing or specialized valve, while the gutter, which is placed at or near the bottom of the shelf, collects refuse water running from tank drain overflows. Refuse water runs from the gutters into a plastic sump or sumps located at the bottom of the rack, where it is then filtered directly or pumped elsewhere for filtration.

There are three basic types of racks. A **modular rack** (also known as a **multi-rack**) is a group of individual racks linked together into one collective system that shares a centralized filtration unit **(Figure 55)**. This is the most commonly used rack type, because of its relatively low cost and enhanced space efficiency. A **standalone rack** is an individual rack that contains a complete filtration unit; hence, it can "stand alone" as a completely self-contained system **(Figure 56)**. Standalone racks are typically used in smaller facilities, or as quarantine racks within larger facilities, or in situations where isolation from other racks or systems is necessary. A **bench-top rack** is a smaller version of standalone rack that is small enough to fit

Fig. 55 Modular rack systems (multi-racks). Photo courtesy of Wendy Porter-Francis.

Fig. 56 Stand-alone rack. Photo courtesy of Wendy Porter-Francis.

on top of the standard laboratory bench (**Figure 57**). These small-volume systems are useful for startup or pilot studies, or in instances where a small population of fish needs to be kept separately outside a larger fish facility.

Water Treatment

The great majority of commercially available zebrafish housing systems employ **recirculating aquaculture** technology. The alternative aquaculture system type, **flow-through** technology, requires

Fig. 57 Bench-top rack. Photo courtesy of Wendy Porter-Francis.

a great deal more space and water than is generally available in research laboratory settings and is used in only a limited number of situations for zebrafish **(Figure 58)**. In recirculating aquaculture systems, the water used to house and grow aquatic animals is treated and reused, thereby greatly reducing water usage and space requirements. The central operating challenge of this type of aquaculture is the removal of the toxic wastes produced by fish metabolism and the breakdown of other organic materials in the system, a process that can be collectively referred to as water treatment. This treatment of the effluent water is the driving force that sustains life in all recirculating zebrafish housing systems; without it, the fish cannot survive.

In a zebrafish housing system, ammonia nitrogen, which is highly toxic, is produced as a result of fish metabolism and the breakdown of organic materials present in the system. During normal system operation, water containing these waste products is flushed from tanks out of overflow drains, collected in system gutters, and subsequently pumped through a sequential series of filters—mechanical,

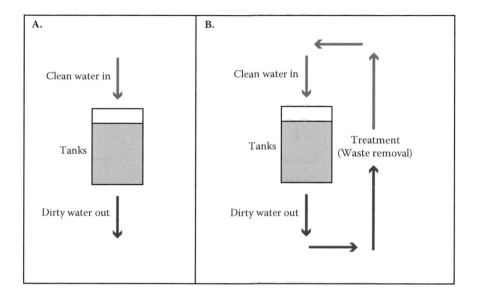

Fig. 58 Flow-through vs. recirculating aquaculture. In flow-through aquaculture (A), clean water (blue arrow) is delivered to tanks and wastewater (red arrow) flows out. In recirculating aquaculture (B), wastewater is treated to remove wastes and then returned to tanks.

biological, and chemical—followed by disinfection. It is during this passage through the filters and the sterilizing units that ammonia, other organic waste products, and pathogens are removed from the water before it is returned as "clean" water to the fish.

Typically, the first step in this process is the sequestering and removal of suspended solids from the effluent water via **mechanical** or **particle filtration**. Solid wastes must be eliminated from the system because they use a significant amount of available oxygen and can produce ammonia. The most common mechanical filter type is a plastic or stainless steel floss, mesh screen, pad, or sponge that traps solids as water flows through it. These filters tend to become clogged over time and must be regularly cleaned to ensure that they are functioning properly.

The second and most important phase of water treatment in a zebrafish housing system is **biological filtration**. Biological filtration (also known as biofiltration) is a term used to describe the oxidation of ammonia to nitrite and then nitrate by certain species of bacteria **(Figure 59)**. These nitrifying bacteria, which are ubiquitous in soil and water, affix themselves to and grow on all surfaces in a fish system. A **biological filter** or **biofilter** is a specially designed

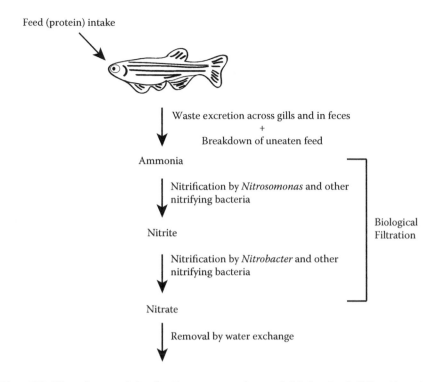

Feed (protein) intake

Waste excretion across gills and in feces
+
Breakdown of uneaten feed

Ammonia

Nitrification by *Nitrosomonas* and other
nitrifying bacteria

Nitrite

Nitrification by *Nitrobacter* and other
nitrifying bacteria

Nitrate

Biological
Filtration

Removal by water exchange

Fig. 59 Simple model of nitrogen cycle and biological filtration in a fish housing system.

substrate with an extremely high surface area upon which these species of bacteria affix themselves to and grow on in very high densities. As effluent water flows through the biofilter, species of bacteria that colonize the substrate, most notably of the genera *Nitrosomonas* and *Nitrobacter*, oxidize the ammonia to nitrite and then the nitrite to nitrate, thereby "detoxifying" the water so that it can be returned to the fish. Nitrate, which is toxic to fish only at very high concentrations, can then be kept at or below maximally acceptable levels by regular water changes.

A number of diverse types of biological filters may be used in zebrafish housing systems **(Table 5)**. These filters vary in substrate type and design, and each has its intrinsic advantages and disadvantages. Among the most common substrates used in the industry are plastic balls, silicon beads, and sponge pads. Available filter designs include fluidized beds, submerged packed filter beds, floating media, rotating biological contactors, and trickling filters. The primary factors determining biological filter efficiency are the surface area of substrate (higher surface areas support higher densities of bacteria

**TABLE 5: COMMON BIOLOGICAL FILTER TYPES,
WITH GENERALIZED SURFACE AREA CLASSIFICATION**

Type	Surface Area
Submerged packed bed	Low
Trickling	Low
Floating media	Moderate
Rotating biological contactor	Low
Fluidized bed	High

and consequently have greater filtering capacity), uniformity of flow rates through the substrate, and dissolved oxygen levels (biological filtration is a highly aerobic process). The precise type and size of the biological filter used must match the density and application of fish being housed in the system. In general, it is advantageous to employ an oversized biological filter in a system so that an increase in the numbers of fish being housed or intensity of management never outstrips capacity.

Another frequently employed step in the water treatment process is **chemical filtration**. While chemical filtration is not considered essential, it is often incorporated in zebrafish housing systems to help reduce colors and odors in effluent water that are not removed during the mechanical or biological filtration phases of wastewater treatment. Perhaps the most commonly employed media in this filter type is activated carbon. This material can be readily purchased from most aquaculture suppliers in granular form, and is often used as a protectant against chlorine, which is toxic to fish and is sometimes present in municipal water supplies. Activated carbon may also be used to bind to and remove other chemicals, such as medications or other molecules used in chemical treatments of the fish, as well as pheromones released by the fish that may interfere with reproduction (Gerlach 2006). Another compound frequently used in similar fashion in chemical filters is zeolite, which is a clay that binds positively charged ions, such as ammonium (NH_4^+). All chemical media used in filters should be well rinsed with water prior to use in systems to prevent clouding of the water.

The final step in the water treatment process is **sterilization** or **disinfection**. Regardless of design, the water in all zebrafish housing systems will contain and support populations of other microorganisms, including bacteria, viruses, protozoa, and fungi, some of which can be pathogenic to fish. Therefore, some degree of sterilization of the water during treatment is necessary to minimize the spread of pathogens that may be present in the system water. Although this process does not eliminate all microbes, it decreases the overall pathogen load in fish housing systems. Of the two primary disinfection techniques used in recirculating systems, ultraviolet (UV) sterilization is the method of choice for zebrafish. The other approach is ozonation, but as it is rarely employed in zebrafish housing systems, it will not be treated here.

All commercially available zebrafish systems incorporate UV sterilizers in design. A typical UV sterilizer consists of a plastic tube that contains a UV bulb sheathed inside a quartz sleeve. During system operation, filtered effluent water passes through the sterilizer tube after it has been through mechanical and biofiltration before it is returned to the fish. As the water flows through the tube, the bulb inside emits UV light at a wavelength that penetrates the cells of microorganisms, destroying their DNA and effectively killing them.

A specific level of UV irradiation, often referred to as a "zap dose," is required to eliminate selected microorganisms. The zap dose is dependent upon a number of factors, including the wattage of the lamp, contact time or flow rate of the water, water clarity, and size and biological characteristics of the target organism (Yanong 2003). In general, smaller and simpler organisms will require lower doses of UV in order to be eliminated, but this is not always the case. Recommended zap doses for a number of important fish pathogens are given in **Figure 60**. To ensure their continued functionality, UV sterilizing units must be maintained regularly according to manufacturer instructions. Such routine maintenance includes regular bulb changes and cleaning of the quartz sleeves.

water source

The water source is the basic foundation of the controlled environment in a zebrafish housing system. While municipal or other local water supplies may be suitable for a variety of everyday uses, they may not be appropriate for fish. There are several reasons for this. Source water may contain chemical and microbial contaminants that can be harmful for fish **(Table 6)**. Also, depending upon local environmental conditions, the basic chemical qualities of the water may not meet the basic preferences of zebrafish. Finally, various qualities of the supply may vary periodically with weather events and season. Given these realities, it is imperative that source water be treated to remove impurities and stabilized in its chemical composition before it is added to housing systems.

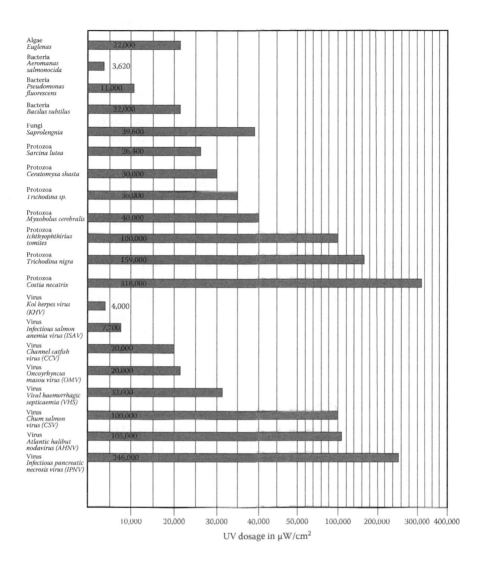

Fig. 60 Required microorganism UV dose (per single pass through UV sterilizer). (Graph courtesy of Emperor Aquatics, Inc., Pottstown, PA).

Testing and Analysis

Ideally, the source water to be used to supply a zebrafish facility should be analyzed for chemical and microbial impurities *before* the facility is constructed. If such testing cannot be conducted in house, most private water testing laboratories are capable of carrying out such comprehensive analysis. Once the contaminant levels of the

TABLE 6: COMMON CONTAMINANTS OF SOURCE WATER

Contaminant	Type	Source
Chlorine	Chemical	Municipal water disinfectant
Chloramine	Chemical	Municipal water disinfectant
Copper	Heavy metal	Pipes, fittings, source
Zinc	Heavy metal	Pipes, fittings, source
Lead	Heavy metal	Pipes, fittings, source
Bacteria	Microorganism	Source
Viruses	Microorganism	Source
Protozoa	Microorganism	Source
Fungi	Microorganism	Source

water source in question are known, the most appropriate treatment method can be selected, usually in consultation with a qualified water treatment expert. Source water should also be periodically tested in the same way, even after the system has been established, so that any changes in quality and/or contaminant levels can be dealt with in a timely and efficient manner.

Treatment Methods

There are a variety of approaches that can be employed to remove undesirable impurities from water. In all cases, it will be necessary to add minerals (via a commercially available aquarium salt mixture) back to the treated water before it can actually be used for housing fish.

Reverse osmosis

Reverse osmosis (RO) is a water purification method whereby high pressure is used to force water through a semi-permeable membrane. The membrane allows pure water (the solvent) to pass through, but rejects any particles that are too large to fit through the holes in the membrane. The RO filtration process effectively removes from water any salts, heavy metals, and virtually any other particles that are larger than the pore size of the membrane. While this includes most bacteria and viruses, it is not recommended that RO be relied on for this purpose, as contamination is possible from periodic degradation of the membrane (Dvorak and Skipton 2008). UV sterilizers should be employed, post-filtration, to eliminate or reduce microbial contaminants.

A number of ions and chemicals, including chlorines and chloramines, which are both highly toxic to fish, are not effectively removed during the RO process. For this reason, RO units should be used in conjunction with activated carbon filters prior to filtration. Activated carbon removes or reduces many organic contaminants, as well as chlorine and chloramines.

Deionization

Deionization is the process of removing salts from water. This is achieved by passing untreated water through a series of ion exchange resin filters, which bind the charged particles in the water and remove them from solution. The resultant deionized (DI) water is free from minerals, but will still contain uncharged organic particles and microbes.

Combined water treatment systems

Since no single water purification method removes all impurities from source water, the best approach is to combine several proven methods within one system to produce the purest possible template. The most common treatment system of this type is a reverse osmosis/deionized (RO/DI) system. In a typical RO/DI system, source water is first passed through a sediment filter to remove any suspended larger particles that may be present in the supply before it is run through a bed of activated carbon to remove dissolved organics. After the carbon filtration step, the water is then doubly "purified," by reverse osmosis followed by deionization. The resultant purified water is then sanitized by UV sterilizers before it is conditioned for fish **(Figure 61)**.

Treatment System Maintenance

Given the importance of maintaining stable, purified template water, it is imperative that treatment systems be serviced regularly according to manufacturer recommendations. In particular, reverse osmosis membranes and ion exchange resins used in deionization units wear down with time and usage, and need to be checked and replaced on a routine basis.

Water Conditioning

Once source water has been treated to remove impurities, it is not technically ready to be used for fish. This is because purified water is necessarily devoid of minerals that fish require for normal metabolism and growth. Therefore, the water must be reconstituted with

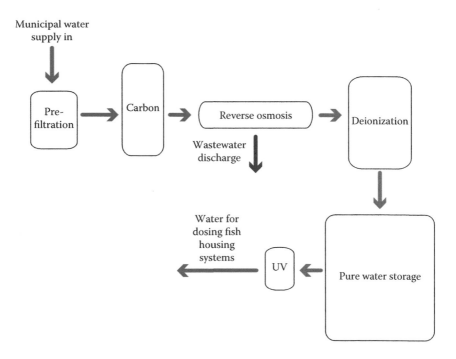

Fig. 61 Flow diagram for an RO/DI water purification system.

salts and minerals, or "conditioned," before it is ready to be distributed to the fish facility and housing systems. The simplest manner in which to condition purified water is to add commercial sea salt mixtures to the water at specific dosage rates that will yield the desired salinity and pH. It is also necessary to add sodium bicarbonate to help increase alkalinity.

Source water can be conditioned in this fashion in one of two ways: either within the system reservoirs, or within off-system reserve storage tanks, prior to distribution to the housing systems. The former approach requires the use of an automatic water quality monitoring system. As pure water enters the system, it will reduce system water pH and salinity. When the monitoring system probes detect that the values of these parameters have fallen below certain predetermined thresholds, they cause the system to be dosed with saline and bicarbonate solutions (from small dosing reservoirs on systems) until the minimum set points are reached. When this method is employed, the regular maintenance and calibration of monitoring system probes becomes absolutely essential, as malfunctioning probes that are not dosing properly will inevitably compromise water quality.

In the alternative method, pure water is dosed with salts and sodium bicarbonate by hand or via an automatic dosing system as it enters an off-system reserve tank. The reserve tank is then used to supply conditioned water to the housing systems on demand. This approach is preferable to in-system conditioning, because the fish do not experience slight fluctuations in water chemistry that occur during the conditioning step. Further, if problems arise during the conditioning process, they are more removed from the housing systems and less of a direct threat to system stability. Finally, the purified, conditioned water can be used for basic tasks in the facility, including for setting up breeding crosses. While this conditioning method is the better choice, it is less frequently employed because of increased costs and space requirements.

water quality management

Along with nutrition and feeding, water quality is the most critical determinant of zebrafish health and productivity. It is well established that water chemistry parameters have a direct impact on zebrafish development, immune function, physiology, anatomy, disease resistance, stress, behavior, and reproduction (Contreras-Sánchez et al. 1998; Groff and Zinkl 1999; Sawant et al. 2001; Evans et al. 2005; van der Meer et al. 2005; Johansen et al. 2006; Craig et al. 2007; Fiess et al. 2007). Further, the efficiency of the biological filtration process is also dependent upon certain elements of water quality. Consequently, the various and interrelated chemical and physical parameters of the water fish are maintained in—including pH, dissolved oxygen, salinity, temperature, nitrogenous wastes, alkalinity, and hardness—need to be controlled and continuously monitored to ensure that they are within the optimal range for zebrafish and proper system function. This is a daunting task that requires not only a thorough understanding of how each component of water quality operates and fits into the overall system, but also an awareness of what values are most appropriate for zebrafish.

Zebrafish are exceptionally tolerant of a wide range of environmental conditions in captivity, reflecting their broad distribution across variable habitat types in the wild. While this trait has played a key role in their rise to prominence as a model, it should be recognized that production and health are at their highest when fish are maintained within an optimum range of environmental parameters.

Simply stated, in terms of quality management, there is a difference between keeping zebrafish alive and keeping them thriving. In addition, water quality can have profound impacts on research.

Key Parameters

Temperature

Temperature exerts profound effects on fish, as well as upon the biological and chemical processes that define their environment. As poikilothermic animals, fish display varying degrees of tolerance to changes in temperature, as well as having a narrower optimum range in which they perform best (Kelsch and Neill 1990). Zebrafish exhibit an acclimated thermal tolerance range of 6.7–41.7°C, which ranks them among the most eurythermal species on record (Cortemeglia and Beitinger 2005). This temperature flexibility is representative of their adaptability to variable conditions, as they are subjected to wide swings in temperature in their natural habitats (Spence et al. 2006).

Under laboratory conditions, zebrafish are typically maintained at 28°C (Matthews et al. 2002; Westerfield 2007). The reasons for this are twofold. First, much of the developmental work done on the species today is based on classic embryological staging studies of zebrafish that were conducted at 28.5°C (Kimmel et al. 1995). Therefore, fish are typically kept at or near these temperatures so that existing standards can be applied. Second, it is apparent that this range is at or near optimum, as growth rates are highest when fish are kept at 28°C (Schaefer and Ryan 2006). Therefore, it is recommended that zebrafish be maintained at stable temperatures within the range of 24–28°C (Matthews et al. 2002).

pH

The relative concentration of acids (hydrogen ions, H^+) and bases (primarily carbonate, CO_3^{2-}, and bicarbonate, HCO_3^-) in a solution determines its pH. One of the central challenges of water quality management in recirculating aquaculture is to maintain water at target pH values in the face of continuous acid production resulting from natural processes in the system **(Table 7)**. Fish respiration, the breakdown of organic wastes (such as uneaten feed), and the metabolism of nitrogenous wastes by microbes in the biological filter all produce acids and decrease pH. Buffers such as sodium bicarbonate or aragonite (a crystalline form of calcium carbonate)

TABLE 7: IMPACTS OF VARIOUS PROCESSES ON pH

Process	Effect on pH
Biological filtration (oxidation of ammonia to nitrate)	Decreases
Decomposition	Decreases
Fish metabolism	Decreases
Aeration	Increases
Water contact with carbon sources (crushed coral, aragonite, sodium bicarbonate, etc.)	Increases
Feeding	Decreases
Photosynthesis	Increases

must be continuously added to the system to prevent pH from falling beneath levels necessary for fish and biological filtration function.

Like temperature, the pH of water in aquatic systems also exerts profound effects on biological processes in fish, as well as on the function of the microbial community that supports them. In closed recirculating aquaculture systems, the optimal pH range for the bacterial flora in biofilms that metabolize nitrogenous wastes excreted by fish is between 7 and 8 (Masser et al. 1999). While most freshwater fish can tolerate a wider pH range of ~6.0–9.5, it is generally practical to maintain most freshwater fish at a pH in the 7–8 range in order to promote good health of the biological filter and stable water quality. However, all fishes display a specific range of preference where growth, feed conversion, and reproduction is optimal; consequently, the goal of pH management in culture is to balance these needs with the requirement of bacteria in biological filters such that production of both is optimized.

Data from field studies suggest that zebrafish are found in slightly alkaline waters (McClure et al. 2006; Spence et al. 2006), although pH values in their habitats can be much lower, especially during the rainy season (Engeszer et al. 2007). This exemplifies their broad tolerance for variations in pH, although wide swings over short time periods are likely to be energetically costly. In laboratory settings, the maintenance pH that most culturing facilities strive for is between 7 and 8, which is within the general range recommended for freshwater culture (Buttner et al. 1993). Managers should strive for stability within this range.

Alkalinity

Alkalinity is the measurement of all bases present in water. Alkalinity can be thought of as "buffering capacity," in that it represents the capacity of the water to resist changes in pH. The most important components of alkalinity, which is typically expressed in mg/L $CaCO_3$, are carbonate (CO_3^-) and bicarbonate (HCO_3^{2-}) ions.

A number of factors tend to produce acids and drive pH values down in recirculating systems **(Table 7)**. As this occurs, the alkalinity level of the water determines the extent to which the pH will actually drop. Well-buffered water is more resistant to change than poorly buffered water. In general, it is advisable to maintain alkalinity values within the range of 50–150 mg/L $CaCO_3$ for fish culture. Levels less than 20 mg/L are considered dangerously low (Wurts 2002).

Alkalinity is also crucially important for efficient biological filtration, as the microbes that populate the filter require the bicarbonate portion of alkalinity for survival and growth (Yanong 2003). If alkalinity values are too low, the populations of these bacteria may crash, resulting in an ammonia spike that could seriously threaten the health of the fish in the system.

In practice, it is important to consider the relationship between alkalinity and biomass in the system. Under high loading situations (high fish densities and intensive feeding regimes), alkalinity values have to be on the higher end of the recommended scale to maintain target pH values. As loading (and biomass) decreases, the requirement for alkalinity is also reduced. Typical zebrafish housing systems are operated at a low biomass relative to the actual capacity of the system, so it takes less alkalinity to maintain pH. Generally, if pH values are stable, then it is likely that alkalinity levels are sufficient. However, alkalinity should be monitored on a regular basis so that managers can react quickly if rapid or unexpected swings in pH values occur.

Aquarium salts used in fish water mixtures will usually contribute some alkalinity to the water, but this is insufficient to maintain pH unless fish densities are extremely low. In nearly all cases, a buffering agent must be added to the water to prevent the pH from dropping below acceptable levels. The most commonly used and readily available buffers are sodium bicarbonate (baking soda) and crushed coral/aragonite beds (see below).

Hardness

Water hardness is a measure of the quantity of divalent ions, primarily calcium and magnesium, and to a lesser extent iron and selenium, in water (Wurts 2002). Fish require these ions for biological function, and they must be provided to fish in culture in the water and/or the diet. The most important of these ions, calcium, is required by fish for ossification, blood clotting, and a number of other biological and physiological processes (Wurts 1993). The degree of hardness may also affect osmoregulation and is often, but not always, related to the buffering capacity of the water. Finally, hardness may also influence the pathology of certain diseases (Canadian Council on Animal Care 2005).

Municipal water supplies will have varying degrees of hardness, depending on the local environment. When deionized or reverse osmosis water is utilized as a source, the water will contain minimal hardness, even after reconstitution with aquarium salts. The hardness of the water may be elevated in a number of ways, including via the direct addition of calcium and magnesium salts to the water.

Hardness may also be increased by passing system water through a crystalline form of calcium carbonate ($CaCO_3$), such as crushed coral, oyster shell, or aragonite. When water comes in contact with one of these substrates, it dissolves the calcium carbonate into its constituent elements, calcium and carbonate. This process, which is both gradual and continuous, simultaneously contributes to hardness and alkalinity of the water **(Figure 62)**.

The simplest way to employ this approach is to add a tank full of coral or aragonite to the system. However, since this method works best when water contact with the substrate is maximized, fluidizing a coral/aragonite mixture in a special filter is more effective. Regardless of setup, the substrate will periodically need to be agitated and/or replaced, as the calcium carbonate in contact with the water will become "used up" over time. Hardness levels should be monitored on a routine basis so that such adjustments can be made accordingly.

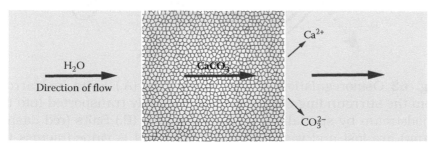

Fig. 62 Aragonite action. When water passing through a bed or column of aragonite drops below a pH of 7.8, the aragonite dissolves and releases Ca^{2+} and CO_3^- into the water, simultaneously contributing to both hardness and alkalinity.

Zebrafish have been classified as a "hard water" species, preferring hardness values in excess of 100 mg/L (Nusslein-Volhard and Dahm 2002). This classification is supported only by limited experimental evidence (Chen et al. 2003) and inferences made from features of habitats in their native range (Lawrence 2007). However, given the well-documented importance of the various components of hardness for fish health and productivity, managers should strive to maintain system levels of at least 75 mg/L CaCO$_3$, the minimum recommended value for general fish culture (Wurts 2002).

Salinity

Salinity is the total concentration of all dissolved ions in water. Freshwater fishes are hyperosmotic to their environment, and thus tend to gain water and lose salts by diffusion across the gills and skin **(Figure 63)**. Consequently, they must maintain their internal water and salt balance by excreting copious amounts of dilute urine while actively transporting ions back into the blood via chloride cells on the gill epithelium.

The energetic cost of osmoregulation varies with the salinity of the external medium, and all freshwater fishes exhibit a preferred level of salinity where this cost is lowest. Maintaining fish above or below

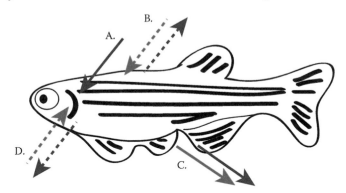

Fig. 63 Osmoregulation in freshwater fish. (A.) Salts (red arrow) from the surrounding environment are actively transported into the bloodstream by specialized cells in the gills. (B.) Salts (red dashed arrow) are lost and water (blue dotted arrow) is gained across the skin via diffusion. (C.) Smaller amounts of salts (red arrow) and large amounts of water (blue arrow) are excreted in urine. (D.) Salts (red dashed arrow) are lost and water (blue dashed arrow) is gained across the gill epithelium via diffusion. Some salts (red arrow) are gained via feed intake.

this optimum is possible (the degree to which depends on the particular species), but because fish must expend more energy in doing so, it can impact growth, survival, and reproduction.

Zebrafish, though a freshwater fish, are tolerant of a wide range of salinities that technically extend to brackish conditions. Embryos tolerate salinities ranging from 0.3 up to 2 g/L (Sawant et al. 2001). This tolerance increases with age, as larval fish may survive 1- to 2-hour pulses of salinities as high as 14 g/L (Sawant et al. 2001). Zebrafish also tolerate very low salinities (less than 0.2 g/L), although long-term maintenance at such levels has a negative impact on egg production (Boisen et al. 2003). These observations are reflective of their adaptation to the variability of salinities in their native habitats, which vary widely depending on underlying geology and seasonal fluctuation in rainfall.

In practice, salinities for zebrafish adults should be maintained at a stable level within the range of 0.5–2 g/L. In some instances, such as in transport and during disease outbreaks, it may be appropriate to keep them at higher salinities, as energy saved as a result of decreased osmoregulatory demand can be allocated toward immune function and the stress response. Increasing salinity may also help reduce the populations of certain fish parasites (Rothen et al. 2002).

Dissolved oxygen

Dissolved oxygen is a critical environmental parameter in aquaculture. Fish require oxygen for respiration, and demand depends upon a number of factors, including body size, feeding rate, activity levels, and temperature. The availability of dissolved oxygen in the water is influenced by elevation, water temperature, and salinity **(Table 8)**.

In general, small-bodied fish such as zebrafish typically have high metabolic rates, and therefore consume by weight proportionally more oxygen than larger fish. This fact, coupled with elevated maintenance temperatures, high densities of fish, and increased levels of feed input typical of intensive zebrafish culture, necessitates that dissolved oxygen levels be maintained at or just under saturation (~7.8 mg/L at 28°C) to ensure continued health of the fish. For informational purposes, the recommended minimum dissolved oxygen level for warm-water fish is 4 mg/L (Wedemeyer 1996), while the minimum required for biological filtration is 2 mg/L (Wheaton et al. 1994).

While levels of dissolved oxygen do not typically change much in stable zebrafish recirculating systems, this parameter should be monitored regularly, and with greater frequency immediately after significant

TABLE 8: THE EFFECT OF VARIOUS SALINITIES AND TEMPERATURES ON THE SATURATION CONCENTRATIONS OF DISSOLVED O_2 (MG/L) IN WATER, AT SEA LEVEL

Temperature °C	Salinity of 0.25 g/L	Salinity of 2.0 g/L	Salinity of 5.0 g/L
20°	9.079	8.986	8.828
25°	8.252	8.170	8.032
28°	7.817	7.741	7.613
30°	7.549	7.476	7.354

changes are made on systems (large increase in number of fish on system, changes in feeding regimen or diet, etc.). However, in situations where tanks are taken off system, dissolved oxygen becomes a much more critical issue to consider, as it can very quickly be depleted to dangerously low levels, especially when densities are higher.

Carbon dioxide

Carbon dioxide (CO_2) is produced in aquatic systems by the respiration of animals, by the photosynthesis of plants and phytoplankton, and during the breakdown of organic matter. It can also be present in water originating from calcareous sources. Levels below 20 mg/L free CO_2 are generally recommended for culture of freshwater fish in recirculating systems (Swann 1997). Excess CO_2 is removed via "off-gassing" by aeration (or by water spilling into sumps or raceways) in recirculating systems, and the use of calcium carbonate ($CaCO_3$) or sodium bicarbonate ($NaHCO_3$) to maintain alkalinity. While free CO_2 is seldom a problem issue in typical zebrafish systems, it is advisable to test this parameter regularly to ensure that systems are operating beneath recommended values.

Nitrogenous wastes

The primary waste product of fish metabolism is ammonia. In freshwater fishes, ammonia is excreted across the gill epithelium via passive diffusion, and to a lesser extent in feces (Wilkie 2002). It is also produced during the decomposition of decaying organic matter (dead fish, uneaten feed). Two forms of ammonia exist in equilibrium aquatic systems, ammonia (NH_3) and ammonium (NH_4^+), the sum of which is referred to as total ammonia nitrogen (TAN). The ratio of the highly toxic NH_3 to the non-toxic ammonium NH_4^+ increases with pH and temperature, and to a lesser extent decreases as salinity is increased **(Table 9)**.

TABLE 9: THE EFFECTS OF pH AND
TEMPERATURE ON THE PERCENTAGE OF
FREE AMMONIA (AS NH$_3$) IN FRESHWATER

pH	20°C	25°C	30°C
6.5	0.13	0.18	0.25
7.0	0.40	0.55	0.80
7.5	1.24	1.73	2.48
8.0	3.83	5.28	7.45
8.5	11.18	14.97	20.29

Levels of NH$_3$ in excess of 0.02 mg/L are generally toxic to aquatic animals, and therefore must be eliminated in closed recirculating systems. This is accomplished in zebrafish systems by nitrifying bacteria that oxidize NH$_3$/NH$_4$ into nitrate (NO$_3$). The intermediate product of this conversion, nitrite (NO$_2$), is also toxic to fish, and can be problematic in freshwater systems at concentrations in excess of 1 mg/L (Wheaton 2002).

The final product of the oxidization process, nitrate, is relatively non-toxic to fish, as most species will tolerate up to 1000 mg/L (Wheaton 2002). In recirculating systems, excess nitrates are most easily removed by periodic water exchanges.

In practice, TAN and NO$_2$ should be maintained at zero levels in zebrafish aquaculture systems, owing to their toxicity at low concentrations. Zebrafish may tolerate chronic exposure to these metabolites, but this is a long-term stressor that should be avoided at all costs. Because it is relatively harmless, NO$_3$ can be present, although it is prudent to keep its levels low to improve overall water quality and to reduce the rate of algal and cyanobacterial growth on tank surfaces. The levels of all three products should be tested on a regular basis, with increased frequency for new systems or for systems that have experienced a change in feeding regime or an increase in the number of fish.

In new systems, it can take up to 6 to 8 weeks to establish bacterial populations on the biological filter when water temperature is 25–28°C (Yanong 2003). The best way to "seed" a new biological filter is to very slowly add small numbers fish to the system over a period of several months. It is also possible to inoculate the system with bacteria obtained from commercial suppliers or from an established, pathogen-free source system (Yanong 2003). In any case, it is imperative to monitor the levels of all three products on a continuous basis until the system stabilizes **(Figure 64)**.

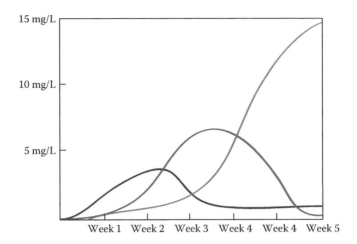

Fig. 64 Biological filter establishment curves. Generalized schema of the concentrations of total ammonia nitrogen (red), nitrite (blue), and nitrate (green) in a newly established system over time. After Timmons, M. B. et al. eds. 2002. *Recirculating Aquaculture Systems*, 2nd ed. Ithaca, NY: Cayuga Aqua Ventures.

Testing

Given the critical importance of water quality for fish health and productivity, the various parameters listed above should be tested and recorded on a regular basis. The frequency of testing will depend on a number of factors. Parameters that are more likely to move more quickly than others, such as pH, salinity, and temperature, should be monitored more frequently, whereas others that tend to be more stable, such as hardness, alkalinity, and dissolved oxygen (in typical zebrafish systems at least), can be tested less often. The frequency of testing also will be dependent upon the age of the system. For example, during system startup, the levels of nitrogenous wastes products (TAN, NO_2, and NO_3) should be tested on a daily basis while the nitrifying bacteria in the biological filter become established. In stable, mature systems, they can be tested on a weekly basis. Finally, it is also prudent to increase the frequency of testing whenever a large change takes place in the system. An example of such a change might be the introduction or removal of large numbers of new fish, the addition of a new substrate or water additive, or an alteration in feeding regime.

Whenever possible, redundancy in water testing methods should be employed. For example, many zebrafish housing systems will have

TABLE 10: WATER TESTING: RECOMMENDED PARAMETERS, TARGET RANGES, AND TESTING FREQUENCIES

Parameter	Target Range	Testing Frequencies
pH	Stable within 6.8–8.5	Daily
Temperature	Stable within 24–28°C	Daily
Total ammonia nitrogen (TAN)	Zero	Weekly; daily at system startup
Nitrite	Zero	Weekly; daily at system startup
Nitrate	Up to 200 ppm (mg/L)	Weekly
Alkalinity	Stable within 50–150 ppm (mg/L)	Monthly
Hardness	Stable within 75–200 ppm (mg/L)	Monthly
Salinity	Stable within 0.5–2.0 ppt (g/L)	Daily
Dissolved oxygen	No less than 4 ppm (mg/L)	Monthly
Carbon dioxide	No more than 20 ppm (mg/L)	Monthly

the option of being outfitted with electronic monitoring systems that will continuously monitor and record selected parameters, including pH, conductivity, salinity, dissolved oxygen, and temperature. These systems are an excellent way to keep track of and in some cases control water quality, but are not foolproof. The probes should be serviced and calibrated regularly according to manufacturer instructions, and recorded values should be periodically cross-checked against colorimetric-type testing kits to ensure that they are within the proper range. This is particularly critical when monitoring systems are being used to dose and control system water pH and salinity. If the probes aren't measuring accurately, they aren't dosing and controlling properly either! A list of suggested parameters, target values, and testing frequencies is given in **Table 10**.

effluent management

Zebrafish facilities are water-intensive entities, consuming and discharging relatively large amounts of water during the course of normal operations. In terms of discharge, while it is mostly dilute, the effluent emanating from a fish facility may contain pathogens, fish, and elevated levels of specific nutrients. Nutrient loading is generally more of a challenge in production aquaculture as a result of the much higher biomass and intensive feeding rates associated with those settings. However, the pathogens and fish that may be contained in zebrafish facility wastewater warrant special attention.

First and foremost, all waste or effluent water discharged from facilities needs to be handled in accordance with local and governmental regulations. The planning for this is typically dealt with by architects at the design phase of a new room, building, or renovation, but should also be considered when a smaller facility is set up within existing space with few or no modifications. Regardless of the size and scope of the operation, wastewater-handling capabilities *must* meet these minimal standards.

There are various ways to treat effluent, depending on the situation. For basic operations (growing and breeding fish for experiments with minimal chemical treatments or exposure to biological agents), the simplest way is not to treat it at all prior to direct discharge into municipal sewer systems. The underlying assumption here is that as long as the municipality has the capacity to receive the volumes being discharged, the water can be adequately treated (Canadian Council on Animal Care 2005).

A further layer of treatment may be provided at the level of the building that contains a zebrafish facility. For example, laboratory buildings at many research institutions in the United States are outfitted with acid neutralization tanks that collect and decontaminate all wastewater from laboratory operations, including zebrafish facilities. Such systems have the capability of effectively handling typical zebrafish facility effluent, and will provide an insurmountable barrier to any live fish or embryos that may find their way into drains.

Finally, it is also possible to reduce the risk of pathogen release into the environment by treating wastewater with UV sterilizers or ozone before discharge into municipal sewage systems. This may be required especially for lines of research that involve the exposure of fish to infectious agents. In these instances, special discharge holding containment units may be required to ensure effective quarantine of these agents.

management

---

regulatory agencies and compliance

Federal agencies overseeing the use of animal research in the United States of America include the Department of Health and Human Services (HHS) and the U.S. Department of Agriculture (USDA). The Animal Welfare Act (AWA), a U.S. federal law, is enforced by the USDA's Animal and Plant Health Inspection Service (APHIS), whose responsibilities include monitoring that research facilities register with the USDA and conducting unannounced yearly inspections of the facilities.

The AWA does not regulate the use of zebrafish. The AWA regulates the use of species that include any live or dead dog, cat, nonhuman primate, guinea pig, hamster, rabbit, aquatic mammal, or any other warm-blooded animal that is being used or is intended for use in research, teaching, testing, experimentation, or exhibition, or as a pet. Birds, rats of the genus *Rattus*, and mice of the genus *Mus* bred for use in research, teaching, or testing, as well as horse and farm animals intended for use as food or fiber or used in studies to improve production and quality of food and fiber, are specifically excluded. In general, the AWA covers warm-blooded animals, and not cold-blooded vertebrates such as fish.

The National Institutes of Health (NIH) is an agency belonging to the United States Department of Health and Human Services (DHHS). The Health Research Extension Act mandated the DHHS to establish guidelines for the care and treatment of animals used in research. The NIH develops many policies within the DHHS; the most

121

pertinent for zebrafish is the Public Health Service Policy on Humane Care and Use of Laboratory Animals (the PHS Policy). The PHS Policy regulates the use, actual or intended, of any live vertebrate animal in research, training, experimentation, or biological testing or for related purposes. This includes zebrafish, and applies to any vertebrate research performed or sponsored by the PHS. Local regulations in some communities may also be required.

The PHS does not provide specific guidance on the care of fish, but it does provide guidance on the use of live embryonated eggs from avian species and other egg-laying vertebrates that develop backbones prior to hatching. These guidelines that were developed for avian species have been extended and applied to the use of zebrafish embryos in research. Recently, the NIH submitted a final report in 2009 to OLAW on the euthanasia of zebrafish (http://oacu.od.nih.gov/ARAC/documents/Zebrafish.pdf), which is the first set of guidelines specific to zebrafish that has been issued by a U.S. regulatory agency.

This guideline states the following: "Current OLAW interpretation of PHS policy considers aquatic species as "live, vertebrate animals" at hatching. Although this is an imprecise stage for zebrafish it can be approximated at 72 hours post fertilization. For purposes of accountability, all stages of development greater than three days of age should be described in an approved Animal Study Proposal. Thus an estimate of the number of larval zebrafish from day 4–7 dpf should be included in Animal Requirements (Section B in the NIH ASP form).

"Since these early stages (4–7 dpf) do not to feel pain or distress, it is preferable that their numbers be separated from zebrafish ≥ 8 dpf. This number can be listed as Column C in the Pain and Distress Category (Section H) of zebrafish ASPs as a separate number from zebrafish ≥ 8 dpf.

"The pain and distress categorization of the ≥ 8 dpf fish should be determined by the investigator based on the specific procedures described in the protocol. The number of animals used may need to be provided as an estimate, particularly with these young larvae, considering their size and normal housing conditions. Estimated numbers may still be used after they have matured to adults if they are group housed."

institutional animal care and use committees

Institutional Animal Care and Use Committees (IACUC) are federally required committees overseeing animal care and use programs, and they function as one of the primary instruments of animal welfare oversight in the United States (Silverman et al. 2006). Regulatory agencies have charged the IACUC with complying with the AWA and PHS Policy to assure animal welfare in research settings.

While fish are excluded from the AWA, they are covered by the PHS Policy, and therefore any activities related to their use in federally funded research requires IACUC review. Although the administrative processes pertaining to the IACUC may differ between organizations, they all must follow the principles and guidelines established by the AWA and PHS Policy, which are fundamental to humane care of all animals (Borski and Hodson 2003).

The IACUC has a wide range of responsibilities, many of which are beyond the scope of this book. However, some of their fundamental responsibilities include (1) evaluating the animal care and use program; (2) conducting semiannual evaluations and inspecting fish facilities and fish procedure areas; (3) reviewing the animal care and use protocols; (4) ensuring that adequate veterinary care is provided; (5) ensuring that an adequate human occupational health and safety program is available; (6) developing annual reports for the responsible institutional or federal official; and (7) ensuring animal welfare.

IACUC protocol review and semiannual inspections are critical components in ensuring fish welfare and promoting good research. The *Guidelines for the Use of Fishes in Research* was written as a supplement to the *Guide for the Care and Use of Laboratory Animals* to provide guidelines to optimize experimental design and procedures while ensuring humane treatment of fish. This is a good resource that describes the roles and responsibilities and provides good background information needed by the IACUC to review protocols. The *Guidelines for the Use of Fishes in Research* does not specifically address zebrafish, but all fish in general. Another useful resource is provided by the Canadian Council on Animal Care's *Guidelines on the Care and Use of Fish in Research, Teaching, and Testing* (Canadian Council of Animal Care 2005). These guidelines provide significant details on the care and use of fish in research. While these guidelines have not been endorsed by the U.S. regulatory agencies overseeing biomedical research involving fish, the overall premises of the CCAC guidelines are generally in line with their goals and mission.

Table 11 lists key components of animal care and use that can be evaluated when performing a semiannual inspection or when reviewing an IACUC protocol involving zebrafish.

TABLE 11: KEY COMPONENTS OF AN ADEQUATE ZEBRAFISH CARE AND USE PROGRAM

Procurement and Source of the Animals
Transportation of Zebrafish and Acclimation
Zebrafish Number Justification and Animal Census
Fish Identification and Tracking
Physical Facilities and Housing
Husbandry Practices
Sanitization Program
Preventive Maintenance Program
Water
Source
Quality Maintenance
Management of Effluent
Housing Density
Feed and Feeding
Diet
Feeding Frequency
Storage
Animal Health Standards
Sentinel Program
Quarantine Program
Training and Training Records
Stress, Pain, and Distress
Occupational Health and Safety Program
Zoonotic Diseases
Preventive Program – Physical Injuries
Fish Restraint
Restraint Device
Anesthetics and Chemicals
Surgical Procedures
Survival Surgeries and Aseptic Techniques
Pre- and Post-surgical Management
Acceptable Research End Points
Euthanasia
Storage and Disposition

Procurement and the Source of Animals

There are significant differences in the availability and standards between procuring rodents and zebrafish intended for use in research. When purchasing rodents from an established and recognized vendor, researchers are assured of receiving animals derived from breeding programs that manage populations such that the rodents have genetic uniformity or variability, have a defined health status, and have been maintained in accordance with strict biosecurity measures and standardized processes. Currently, this scenario does not apply to the laboratory zebrafish. The few vendors that do supply zebrafish for research should be evaluated on the basis of their genetic management and health-monitoring programs prior to purchasing fish.

Laboratory zebrafish should meet a minimum standard appropriate for research applications that have been established and defined in other biomedical research models such as mice and rats. These standards should provide the basic framework for creating a model that enables reproducible research. The production and management of laboratory zebrafish should not be a source of variability leading to potential research artifacts. It is important to prevent the introduction of non-protocol sources of variation originating from microbial organisms, genetics, water contaminants, and nutrition. Depending on the line, the genetics, morphology, physiology, and behavior of the laboratory zebrafish should be standardized and predictable.

Commercial suppliers that currently sell zebrafish for research service mainly the aquarium trade and grow the fish in greenhouses or outdoor ponds. There are great differences between pond fish and laboratory-reared fish. Fish reared outdoors may present a real biosecurity risk to research populations through the introduction of foreign pathogens. Most, if not all, commercial suppliers do not have programs in place to strictly monitor genetic background, microbial status, and environmental parameters. It is imperative to determine whether the selected fish are suitable for the proposed type of research. Currently, the Zebrafish International Resource Center (ZIRC) is the only commercial supplier of laboratory-reared wild-type, mutant, and transgenic zebrafish.

occupational health and zoonotic diseases

Occupational health programs are responsible for providing recommendations to promote a safe work environment by establishing preventive measures to minimize the risk of disease transmission and the risk of physical injury while working in an aquatic animal research facility.

Zoonoses are diseases of animals that are transmissible to humans. There are several aquatic microorganisms, including viruses, bacteria, fungi, parasites, and dinoflagellates, known to cause diseases in humans. However, the vast majority of these illnesses are secondary to food consumption (Nemetz and Shotts 1993; Lehane and Rawlin 2000; Lowry and Smith 2007). Overall, the incidence of disease transmission between fish and humans is low (Lowry and Smith 2007). In a laboratory setting the most common routes of transmissions include inadvertent ingestion of contaminated aquarium water, secondary infections of open wounds, and direct contact with an infected fish. Many bacterial aquatic organisms that infect humans are opportunistic, and the development of disease in humans may require a preexisting condition where the individual may be immunosuppressed from medication, illness, or pregnancy. However, the number of organisms that realistically pose a risk to human health while working with zebrafish in a research setting seems small. It is important to note that to date the known microorganisms that most commonly affect fish handlers are almost entirely bacteria. In a research setting working with zebrafish, the potential list is even smaller. Thus, zoonotic fungi, viruses, and parasites will not be discussed in this section; for readers interested in learning more regarding that topic, there are several good reference articles that can be reviewed (Nemetz and Shotts 1993; Hoole et al. 2001).

The following aquatic microorganisms may be associated with topically acquired bacterial zoonoses from zebrafish. This list was tailored to include bacteria that have been reported in laboratory zebrafish or in other cyprinids. Common bacteria such as *Aeromonas* spp., *Pseudomonas* spp., and *Mycobacterium* spp. are known to infect zebrafish and can pose a risk to humans (Hoole et al. 2001; Rawls et al. 2004; Clatworthy et al. 2009). Currently, there are no published clinical reports of zoonotic transmission from zebrafish to humans; however, zebrafish are potential hosts to these ubiquitous organisms that have been associated with human disease (Rawls et al. 2004; Cipriano 2001; Rodríguez et al. 2008).

There are several aeromonads associated with human disease that are found in finfish and cause clinical disease in freshwater fish housed in cultured environments (Lowry and Smith 2007). Aeromonads, found throughout the world, commonly cause disease in wild and captive fish and have been identified in several warm- and cold-water species (Cipriano 2001). These include *Aeromonas hydrophila, A. caviae, A. sobria,* and *A. schubertii* (Palumbo et al. 1989; Lowry and Smith 2007). The only published research reporting a natural infection in zebrafish involved *A. hydrophila* (Pullium et al. 1999). *Aeromonas hydrophila* is a facultative anaerobic Gram-negative bacillus commonly found in freshwater environments. It is a normal bacterium found in the gastrointestinal tract of fish and is known to cause ulcerative skin lesions and death in zebrafish (Pullium et al. 1999). It can infect skin lesions in fish and humans. Immunocompetent individuals can develop an infection; individuals with a suppressed immune system are much more susceptible. Potential clinical signs in humans with infections include gastroenteritis and localized wound infections.

Streptococcus iniae, a ubiquitous Gram-positive non-motile beta-hemolytic cocci, has been associated with disease outbreaks in freshwater and saltwater fish, most commonly infecting freshwater fish in recirculating systems (Shoemaker and Klesius 1997; Kusuda and Kasadi 1999; Lehane and Rawlin 2000; Colorni 2002). Infections occur as a consequence of overcrowding and poor water quality. *Streptococcus iniae* has been isolated from zebrafish (*Danio rerio*), pearl danio (*D. albolineatus*), tetras (*Hyphessobrycon* spp.), tilapia, rainbow trout, and others (Ferguson et al. 1994; Shoemaker and Klesius 1997; Lehane and Rawlin 2000; Russo et al. 2006). Zebrafish are very sensitive to this bacterium, and adult zebrafish are used as a research model for streptococcal infection (Phelps et al. 2009; Neely et al. 2002). Human infections have been reported in individuals who injured themselves while handling contaminated water or infected fish. Clinical signs may include cellulitis, and in rare severe cases individuals may develop endocarditis and meningitis (Weinstein et al. 1997; Lehane and Rawlin 2000; Lowry and Smith 2007).

Pseudomonas aeruginosa is a ubiquitous Gram-negative bacterium found in water and soil. It is one of the most common causes of nosocomial infections in the United States, infecting immunocompromised individuals (Clatworthy et al. 2009). *Pseudomonas aeruginosa* has also been found in drinking water in rodent research facilities (Jacoby et al. 2002). It is plausible that depending on the water source, laboratory zebrafish may also be exposed to this opportunistic agent. Again, it is important to note that to date, there are no published reports of naturally occurring infections in zebrafish, but zebrafish are used to study *Pseudomonas* infections (Clatworthy et al. 2009).

Organisms causing atypical mycobacteriosis belong to the genus *Mycobacterium*. They are ubiquitous non-motile Gram-positive acid-fast rods. Mycobacteriosis has been reported to affect a wide range of freshwater and marine fish. The most common forms of mycobacteria include *M. fortuitum*, *M. marinum*, and *M. chelonae*. These three species are the etiological agents of fish mycobacteriosis and can cause zoonotic infections commonly referred to as fish tank granuloma. Infections occur when people are in contact with contaminated water from aquariums and fish tanks or while handling, cleaning, or processing infected fish. Frequently atypical mycobacteriosis may manifest as a single cutaneous granulomatous nodule or non-healing ulcer on the hand or finger. Workers who are immunocompromised or have skin lesions are most at risk. Symptoms may develop weeks to months after exposure.

The best-known practice for preventing transmission of diseases from fish to humans while contacting them and the water is to wear latex gloves, especially if skin wounds are present. Many of the high-risk bacteria are ubiquitous and are very likely to be found in any well-established recirculating system. Another good practice is to wash hands and forearms after handling fish and aquarium water. If someone is injured while working in the fish facility, the incident should be reported and institutional guidelines should be followed. Minor cuts and abrasions should be immediately rinsed and washed with soap and then protected from exposure to fish and aquarium water.

The risk of introducing any microorganism in a zebrafish facility is theoretically greater when zebrafish are introduced from ponds rather than from a population with an established health status or are introduced as bleached embryos. Compared to laboratory-reared zebrafish, fish reared in ponds are exposed to a much greater range of microorganisms from a variety of sources, such as soil, vegetation, aquatic and avian wildlife, and other fish species reared in adjacent ponds. Not only does this increase the risk of introducing unwanted organisms into the facility and potentially exerting negative impacts on husbandry, fish health, and potentially human health, it can also realistically affect research data. Although particular microorganisms may or may not always cause observable clinical signs in fish, it is clear that bacteria in zebrafish have a wide range of impacts, including affecting toll-like receptor-signaling pathways, cytokines and chemokines, immune responses, maternally transferred immunity, and the complement system (Sitjà-Bobadilla 2008; Wang et al. 2009; Sullivan and Kim 2008; Stockhammer et al. 2009). Although these findings were collected from inoculated animals as opposed to natural infections, caution should be taken regarding the potential impact of commensal or opportunistic microorganisms harbored by research zebrafish.

Microorganisms such as *Edwardsiella*, *Escherichia*, *Salmonella*, and *Klebsiella* are found in freshwater environments (Lowry and Smith 2007; Stockhammer et al. 2009). Although no naturally occurring zebrafish infections or zoonotic risk has been documented between laboratory zebrafish and humans, these bacteria have a known zoonotic potential and have been directly associated with fish infections or freshwater aquatic environments. Furthermore, the laboratory zebrafish is used as a model to study infectious diseases and immune function for all the above organisms (Petrie-Hanson et al. 2007; Lowry and Smith 2007; Sullivan and Kim 2008; Stockhammer et al. 2009).

In summary, having a reliable source of fish, using effective sanitization agents, wearing latex gloves, washing hands frequently, preventing oral inoculation, not exposing open wounds to aquarium water and fish, implementing a quarantine program, and not introducing fish with a high risk of wild-life microbial microflora are all good preventative measures.

disposal of fish

Laboratory zebrafish should not be released into the wild, and therefore should not be released into bodies of water such as ponds, rivers, or lakes. Dead fish must be disposed of according to federal, state, and municipal regulations.

Upon completion of studies, embryos and adult fish should be euthanized according the *AVMA Guidelines on Euthanasia*. Carcasses and animal tissue should be kept in a separate storage area maintained below 7°C (44.6°F). Animal-carcass and animal-tissue waste refrigerated storage should be separated from other cold storage. Zebrafish tissue and embryos should not be disposed of through the drain system. Carcasses can be incinerated or collected by a licensed contractor. Disposal of zebrafish used in research programs involving infectious agents or hazardous compounds should follow procedures in accordance with the institution's occupational health and safety policies.

veterinary care

basic biologic parameters

Zebrafish are small, active, schooling fish with 5 to 7 bilateral horizontal stripes on either side of their bodies (Spence et al. 2008). They have 3 unpaired fins and 2 paired fins. Zebrafish also have 2 pairs of barbels on the ventral lateral aspect of the oral cavity (Barman 1991; Hansen et al. 2002; Spence et al. 2008)

One-year-old zebrafish can measure between 25 mm to 35 mm in length (Spence et al. 2008). Zebrafish growth rate is fastest during the first 3 months of life and approaches zero by about 18 months of age (Spence et al. 2008; Spence et al. 2007). The growth rate of laboratory-raised zebrafish is greater than the growth rate of wild zebrafish (Spence et al. 2008). Females are larger than male zebrafish regardless of whether they originate from the wild or the laboratory (Eaton and Farley 1974; Spence et al. 2008). Males are more slender than females, have a yellowish cast on their abdomens, and tend to have larger anal fins than do females (**Figures 65 and 66**) (Spence et al. 2008; Laale 1977).

Zebrafish have a short generation time, ranging from 2 to 4 months, and fertilization is external (Spence et al. 2008). In the wild, zebrafish are thought to spawn prior to and during the rainy season (Spence et al. 2008). Under laboratory conditions, zebrafish can spawn daily, producing several hundred eggs or more in a single clutch (Spence et al. 2007; Spence et al. 2008). At fertilization, the eggs measure approximately 0.7 mm in diameter (Spence et al. 2008). The embryos are optically transparent for the first several days postfertilization

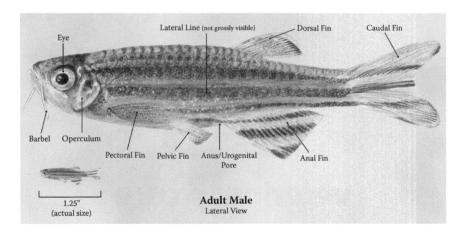

Fig. 65 Adult male zebrafish. Males are slender and may have a yellowish cast on their abdomen. Males may have more pronounced anal fins than females. From Zebrafish Anatomy: *Danio rerio* poster. Copyright 2007, AALAS. Reprinted with permission of AALAS, Inc.

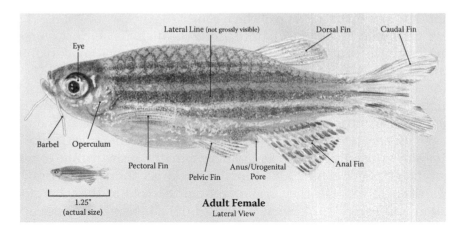

Fig. 66 Adult female zebrafish. Females are larger than males. From Zebrafish Anatomy: *Danio rerio* poster. Copyright 2007, AALAS. Reprinted with permission of AALAS, Inc.

(dpf), and all major organs form within 36 hours postfertilization (hpf). Zebrafish larvae start seeking food and display active avoidance behavior by 5 dpf (Spence et al. 2008).

In nature, zebrafish appear to be primarily an annual species, although some individuals will persist into a second season (Spence et al. 2006). The lifespan of zebrafish in laboratory settings is approximately 42 months, with an upper range of 66 months (Gerhard et al. 2002). Mutant strains may have a shorter life expectancy, depending on the genetic defect(s) present. Skeletal length increases with age up to death, which is suggestive of indeterminate growth (Gerhard et al. 2002).

A common age-related phenotype noted by Gerhard is spinal curvature, secondary to muscle abnormalities (Gerhard et al. 2002). This phenotype has not been observed in wild zebrafish populations (Spence et al. 2008). Since *Pseudoloma neurophilia* infection is the most common disease noted in laboratory zebrafish and infections commonly lead to scoliotic changes due to muscle lesions (Matthews et al. 2001; Kent and Bishop-Stewart 2003; Whipps and Kent 2006), it may be that microsporidiosis is a contributing factor to the pathological development of spinal curvature in aging zebrafish (**Figure 67**).

animal welfare indicators for fish

Animal welfare indicators have been developed for mammalian species. Comparable indicators have not been well established for zebrafish: this is a complex topic where empirical data, ethics,

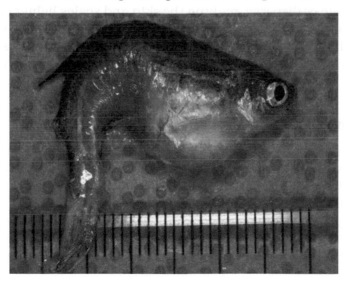

Fig. 67 Zebrafish with scoliotic change and PCR positive for *Pseudoloma neurophilia*. Photo courtesy of Dr. Susan Westmoreland, Assistant Professor of Pathology, Harvard Medical School, Chief, Section of Comparative Pathology.

neuroscience, animal research, and human emotion have resulted in some controversy (Iwama et al. 2004; Johansen et al. 2006; Sørum et al. 2007; Volpato 2009).

The following are proposed indicators for zebrafish well-being, based on those devised for the pet fish and aquaculture industries.

Change in Color

Zebrafish have three types of pigment cells, which are responsible for their unique color pattern: dark-blue melanophores, gold xanthophores, and iridescent iridophores (Spence et al. 2008; Parichy 2006). These pigment cells are responsible for the zebrafish longitudinal stripes extending from the caudal end of the operculum to the caudal fin (**Figure 65**) (Barman 1991; Spence et al. 2008).

Fish change their skin color in response to such stimuli as environmental cues, social subordination, and physiological stress (Höglund et al. 2000, 2002; Pissios et al. 2006). Melanin-concentrating hormone (MCH) is a peptide secreted by the pituitary gland in teleost fish in response to stimuli, and melanosomes are the pigmented granules which are found in melanophores cells located in the scales (Pissios et al. 2006). When MCH is secreted, it causes the melanosomes either to aggregate or disperse, making the skin and scales lighter or darker, respectively (Pissios et al. 2006). It has been shown that the zebrafish melanophores will aggregate or disperse in response to various stimuli, such as camouflage, light intensity, and aggressive behavior (Gerlai et al. 2000; Larson et al. 2006; Guo 2004; Spence et al. 2008).

Respiratory Rate

The respiratory rate of fish can be visually assessed by measuring opercular movement, a means for providing water flow through the gills for oxygenation (Noga 1996). Increased respiratory rate may indicate that the fish is having difficulty breathing (Saha et al.1999). This may be caused by a single or several factors, including decreased water oxygen levels, decreased functional surface area of the gills as a result of disease, and increased physiological demand caused by physical movement.

Swimming and Behavioral Patterns

Swimming behavior, pattern, speed, and position in the water column have been used as welfare indicators in fish, since they are affected by factors such as environmental conditions, stress, predation, and health (Masuda et al. 1993; Kuwadaa et al. 2000; Øverli et al. 2005). Zebrafish are used as a model to study behavior (Gerlai et al. 2000; Lindsay and Vogt 2004; Risner et al. 2006; Gerstner et al. 2009). It is therefore important to determine exogenous factors that may influence zebrafish behavioral research as well as to determine behavioral cues that could be used as welfare indicators.

Growth Rate

Growth rates in fish can vary with genetics, environmental parameters, stocking density, chronic stress, and nutrition (Priestley et al. 2006; Siccardi et al. 2009). Long-term growth rate reduction in fish may indicate chronic stress (Bernier et al. 2004; Pinto et al. 2007). In mammals, the rate of growth slows and stops after reaching sexual maturity. Zebrafish, however, have the capacity for unlimited growth (Gerhard et al. 2002; Kishi 2006). Zebrafish continue to grow after sexual maturation, with the growth rate slowly decreasing at later stages of life; and population density, body size, and age may influence growth rate in zebrafish (Tsai et al. 2007). Conditions for optimal growth and maintenance should be established in order to use growth as an animal health indicator (Siccardi et al. 2009).

Body Condition

Evaluating the body condition of animals such as dogs, cats, cows, and horses is a common indicator used to assess general health. However, body condition scoring is not well established in fish, although various approaches have been proposed (Priestley et al. 2006; Weber and Innis 2007; Cade et al. 2008). The body shape of zebrafish is easily assessed, and changes may be associated with weight gain, weight loss, breeding cycle, non-infectious illnesses, and infections. There are several infectious agents that are known to affect the body condition of zebrafish. Examples include *Pseudoloma neurophilia*, which can lead to body curvature, and atypical *Mycobacterium* spp., which can lead to emaciation (Astrofsky et al. 2000; Matthews et al. 2001; Kent and Bishop-Stewart 2003; Whipps and Kent 2006).

External Morphological Abnormalities

Morphological abnormalities affecting the fins, gills, and eyes are general indicators of well-being, which can be diminished by suboptimal husbandry and environmental conditions. More specifically, frayed fins and gill discoloration are non-specific changes that may occur as consequences of multiple factors; however, they may indicate the presence of a chronic underlying problem associated with poor water quality (Noga 1996).

Reproductive Performance

There are many factors that influence reproductive performance of fish (Lewbart 2002; Schreck 2009). Broad examples may include chronic stress, water chemistry imbalance, urogenital anatomy, pheromones, endocrinology, nutritional imbalance, and age (Contreras-Sánchez et al. 1998; Astrofsky et al. 2000; Lewbart 2002; Brion et al. 2004; Hirai et al. 2006; Mukhi and Patino 2007; Schreck 2009; Alsop et al. 2009).

Biochemical Welfare Indicators

Exposure to stressful conditions can lead to physiological changes that are known to impact energy use, osmoregulation, respiration, and immunity. A fish may adapt to stress for short periods of time when it may behave normally, but it is depleting energy reserves because of the extra requirements placed on it while trying to adapt and compensate for the stressors (Francis-Floyd 2002).

Chronic stress is known to affect fish physiology and morphology, and can potentially impact reproductive performance, behavior, the immune system, and disease resistance (Barton 2002; Dror et al. 2006; Harper 2006; Thomas et al. 2007; Wang et al. 2008; Ramsay et al. 2009; Harper and Wolf 2009). Measuring the stress response in fish has not been standardized, since stress is a continuum and not an all-or-nothing event. In many instances there may be a lack of clinical manifestation, and depending on the magnitude and duration of the exposure, disease may develop over time (Noga 1996; Harper and Wolf 2009). Other challenges include the potential stress associated with sample collection, which may also confound the result (Harper and Wolf 2009). Techniques that have been developed to quantify stress response in a variety of fish include the following:

1. Measurement of the body weight of the fish or measuring specific organs (Dutta et al. 2005; Hosoya et al. 2007; Spencer et al. 2008)

2. Biochemical assays such as for plasma cortisol, corticosterone, glucose, tissue damage enzymes, and heat shock proteins (Barton 2002; Acerete et al. 2004; Dutta et al. 2005; Hosoya et al. 2007; Iwama et al. 2004; Olsen et al. 2008; Trenzado et al. 2008)

3. Assays of immune function (Choi et al. 2007)

4. Assays of gene expression patterns (van der Meer et al. 2005; Marques et al. 2008)

5. Measurement of fish steroids in water and in feces (Turner et al. 2003; Scott and Ellis 2007)

6. Examination of macroscopic and microscopic anatomy (Olsen et al. 2002; Harper and Wolf 2009)

Food Consumption

Food consumption is an important indicator of well-being in many species; it has been proposed as an indicator of well-being in fish, since both acute and chronic stress can impact the appetite of fish. Although the relationship between food consumption and well-being has not been defined in zebrafish, food consumption rates have been studied to assess chemical toxicity, obesity, and bioenergy (Roex et al. 2003; Song and Cone 2007; Chizinski et al. 2008; Siccardi et al. 2009). It is important to note that there are known factors that can lead to variations in individual consumption rates in fish, as well as species-specific differences in how fish respond to stress (McCarthy et al. 1993). It is known that fish decrease their food consumption and mobilize energy reserves in response to stress, and that food-deprived fish are less resistant to stress than fish with sufficient food (Olsen et al. 2008). Although food consumption standards have not been harmonized across the zebrafish research community, establishing laboratory baselines for this behavior can potentially help provide some insight when fluctuations occur.

stress response in fish

Stress results when an animal is exposed to a situation that is outside its realm of tolerance. Common stressors for fish include capture; transport; handling and crowding; stripping; injecting; fin biopsy; sorting; grading; malnutrition; variations in temperature, oxygen, and salinity; environmental contaminants; and pathogens (Øverli et al. 2005; Harper 2006; Dror et al. 2006; Ramsay et al. 2009; Harper and Wolf 2009; Schreck 2009; Siccardi et al. 2009).

Many factors are known to influence animal research in mammalian models (Lipman and Perkins 2002). However, limited attention has been given to the existence of potential artifacts and variables affecting zebrafish bioassays and research modeling (Lipman and Perkins 2002; Harper and Wolf 2009). Some stress responses can be visualized by gross or microscopic examination of various tissues such as gills, liver, and skin (Olsen et al. 2002; Scott and Rogers 2006; Spencer et al. 2008; Harper and Wolf 2009).

The stress response in fish can be divided into three general phases: primary, secondary, and tertiary (Barton 2002). The **primary response** involves a neuroendocrine response in which catecholamines (epinephrine and norepinephrine) are released by chromaffin cells and cortisol is released by the interrenal cells (Barton 2002; Harper and Wolf 2009). The **secondary response** is then initiated, triggering physiologic and metabolic pathways that include hyperglycemia secondary to glycogenolysis and gluconeogenesis; arterial vasodilation in gills; and increased cardiac stroke volume (Gratzek and Reinert 1984; Harper and Wolf 2009). The **tertiary response** involves depression of immune function. Ultimately the adrenal gland releases hormones that decrease the ability of the fish to develop an inflammatory response. Consequently, the fish becomes more susceptible to infection when exposed to a pathogen (Francis-Floyd 2002; Huising et al. 2005). Under stressful conditions, fish redirect their use of energy to adapt to the stress, as opposed to using it for growth or reproduction. The primary and secondary phases are adaptive, and the fish adjusts to the stress while maintaining homeostasis. In contrast, the tertiary phase can lead to systemic changes where the fish does not reach homeostasis and is unable to adapt (Barton 2002; Schreck 2009).

Stress can also affect mineral metabolism in fish. Osmoregulation is an active process where the fish maintains its internal water–salt balance using several organs, including the kidneys, the gills, the alimentary tract, and the urinary bladder (Greenwell et al. 2003). When exposed to continuous stress, the ability of the fish to osmoregulate is affected and forces the fish to utilize more energy than usual to maintain homeostasis (Francis-Floyd 2002).

Anything that impacts the immediate physical environment of the fish, such as water quality, stocking density, light, and sound, can be considered a potential stressor. In a zebrafish facility, poor water quality, including elevated organic load, low dissolved oxygen, high nitrogenous waste levels, and rapid changes in pH and temperature, are important stressors that can precede disease outbreaks (Francis-Floyd 2003; Canadian Council on Animal Care 2005; Lawrence 2007). Water temperature has a direct effect on the physiology, hematology, and clinical chemistry of fish, and rapid decreases in temperature can interfere with the humoral immune response of the fish, decreasing natural killer cell function (Hrubec 1997; Groff and Zinkl 1999; Fiess et al. 2007).

clinical case management and sample collection

The clinical presentation of disease in fish is often the result of a complex interaction of factors. Therefore, when a clinical case is presented, it is necessary to gather as much information pertaining to the affected animals as possible. Some of the most critical factors are discussed below.

Source of the Fish

The source of the affected fish should be identified to determine whether it originated internally, from an outside laboratory, or from a commercial vendor. Most zebrafish that are imported into research laboratories typically originate from other research facilities or from the primary academic supplier, the Zebrafish International Resource Center (ZIRC). Less typically, fish may also be imported from commercial suppliers that grow fish outdoors in ponds for the pet trade. Knowing the source of the fish will help characterize the health status of the animals at the time they entered the facility as well as determine whether the outbreak is localized to a particular fish, tank, life support system, or facility.

Management Practices

It is important to determine the kind of quarantine program that was used at the source, since practices can vary greatly between institutions. Some laboratories have no quarantine program, others have a quarantine rack in the zebrafish holding room, and others may maintain a quarantine rack that is separate from the holding room. The last is the best scenario since there is a physical separation between the quarantine and regular housing, thus reducing the risk of cross-contamination. Additionally, some programs will house the adult fish in quarantine for a defined period of time and then introduce the fish into the holding room. In other programs the quarantined fish never enter the zebrafish housing rooms, with only their bleached embryos being transferred into the housing systems. This is the ideal approach to prevent the inadvertent introduction of a pathogen into a holding facility, since a comprehensive health evaluation is not possible owing to the lack of available diagnostics for many diseases of zebrafish. Although allowing only bleached embryos into facilities is the most effective strategy, it is not likely to prevent the transmission of all pathogens into a facility. For example, the protocol presently used by most zebrafish research facilities to surface disinfect embryos (Westerfield 2007) is ineffective at eliminating *Pseudoloma neurophilia*, one of the most prevalent pathogens in zebrafish populations (Ferguson et al. 2007).

Genetic Background

The genetic background of the animals should be identified. For example, mutant fish and transgenic strains may possess unknown phenotypic characteristics or other physiologic or immunologic factors that may directly impact the immune system and health of the animals.

Age of the Fish

Age has an impact on overall resistance to pathogens. Larval forms and older animals are usually more susceptible. The immune function of an animal also tends to decrease with age, so older fish are more likely to show clinical signs of disease.

Research Application

Establishing previous experimental use of the affected fish is important since they may become susceptible to opportunistic or commensal organisms that under normal circumstances would not cause disease. For example, any procedures that would require elevated levels of handling or disturbance, or exposure to chemicals, could be causative contributing factors in the clinical presentation of disease.

Water

It is important to evaluate water parameters as soon as the fish are demonstrating clinical signs and before water changes are performed. Water toxicity and environmental imbalances can be transient and may cause a range of abnormal behaviors. The diagnosis is potentially easy to miss if the water samples are not collected as soon as the fish demonstrate clinical signs.

Water samples should be collected using a clean (devoid of soap residue) and sealed container. Parameters that should be evaluated include, but are not limited to, ammonia, nitrate, nitrite, chlorine, dissolved oxygen, temperature, pH, and conductivity. Dissolved oxygen and temperature should be measured immediately since these parameters will change over a short period of time. Usually, severe and especially sudden water chemistry imbalances lead to acute clinical disease that affects fish from many tanks as opposed to a few tanks. Recent water chemistry and colony mortality logs should be reviewed to identify trends.

Physical Examination

It is vital that fish care personnel develop a familiarity with the behavior and physical characteristics of normal fish so that they are able to readily identify problem situations when they occur. Some abnormal behavioral and physical changes are described below:

1. Zebrafish are schooling animals and usually stay together. A fish that is consistently separated from the others may indicate an underlying problem.

2. Buoyancy control enables fish to swim throughout the water column in a controlled and purposeful manner. Loss of control is noted when a fish is unable to maintain its position in the water column, resulting in the affected animal either sinking to the bottom of the tank or floating to the surface. The fish may also display repeated attempts to swim in the opposite direction without success.

3. The respiratory rate of fish can be quantified by counting the number of opercular movements. An increased respiratory rate may be an indication of either a water quality problem or a physical impairment of the gills.

4. Lethargic animals may show reduced activity compared to "normal," active animals. Lethargic animals may be "hanging" weakly in the water column, or staying at the bottom or toward the very top of the tank.

5. A group of fish at the surface of the water gasping for air is referred to as fish "piping" for air. This may be caused by supersaturation of the water or by hypoxia, and this behavior would be noted in every tank, or across many tanks, in one system (Noga 1996).

6. Decreased appetite may be observed. Healthy zebrafish will feed actively when presented with palatable food items. When fish do not display this behavior, it may be a sign that something is wrong.

7. Flashing (quick twisting swimming motions and rubbing against surfaces) is commonly seen in fish infected with ectoparasites.

8. Character and color of scales, skin, fins, and gills are significant. Fish with frayed fins is a common indicator of poor water quality, while skin lesions indicate possible bacterial involvement.

9. Excess mucus on the skin may be observed when the fish are exposed to a chronic irritant such as parasites or an unfavorable water chemistry parameter.

10. Skeletal deformities in zebrafish are common findings and have been associated with nutritional deficiencies in fish, microsporidian infections, and advanced age (Matthews et al. 2001; Whipps and Kent 2006; Gerhard et al. 2002).

11. The overall body condition of the fish and its abdominal shape may indicate whether the fish is anorexic, egg bound, or affected by a pathogen. This is determined by seeing whether the abdomen is concave or convex relative to the ventral aspect of the operculum.

12. Eye coloration and size should be the same. Exophthalmos is the forward protrusion of one or two eyes. There are several conditions that result in exophthalmos in fish, but sepsis is typically the primary cause.

For effective diagnostic testing, it is important to use fresh tissue samples, since dead fish rapidly undergo autolysis. Gill tissue and the gastrointestinal tract deteriorate quite quickly, promoting parasitic and bacterial overgrowth. Tissues that have autolyzed should not be used for microbial cultures, parasite evaluation, or histology. Frozen samples can be used for bacterial or viral isolation but not for diagnosing external parasites or performing histology. Fish can be fixed in 10% buffered formalin in preparation for histological evaluations; however, small tissue samples should be used to assure proper fixation.

Once the fish are euthanized, they should be weighed and measured. Physical examination of zebrafish can easily be performed under a dissecting microscope where the character and color of the scales, skin, fins, and gills can be assessed. The presence of lesions, hemorrhagic foci, ulcers, or excessive mucus is easily noted.

Cutaneous wet mount is a rapid diagnostic tool that can be used to identify the presence of lesions, nodules, parasites, and bacterial load of the gills, skin mucus, and fins. It is simple and inexpensive to perform and provides valuable information (Stoskopf 1993; Noga 1996; Astrofsky et al. 2002). The fish should always be handled with wet, non-powdered gloves, since the powder can contaminate the sample and lead to artifacts that may confound analysis of the wet mount under the microscope.

Simple protocol for preparing cutaneous wet mount preparations

Materials needed when performing wet mount preparation include

- MS-222 and sodium bicarbonate for euthanasia
- Glass slide
- Cover slip
- Scalpel blade
- Small or microsurgical scissors
- Small or microsurgical forceps

Place a drop of water in the center of a glass slide and add the tissue sample to be examined (**Figure 68**). The cover slip is then placed on the slide and the sample can be examined under low magnification (4X or 10X). The activity level of parasites may decrease after being exposed to MS-222. Stains can be used to enhance the contrast when assessing cutaneous wet mounts.

Fig. 68 Water droplet in the center of a glass slide.

Gill wet mount preparation

In adult zebrafish, the gills are the major site of gas exchange, acid–base balance, ionic regulation, and excretion of nitrogenous waste (Jonz and Nurse 2005; Rombough 2002; Hwang 2009). Since the gills deteriorate rapidly after euthanasia, they should be the first sample collected and evaluated.

To access the gills, the operculum should be removed by excision using a small pair of scissors (**Figure 69**) (Noga 1996; Astrofsky et al. 2002). A section of the gill lamellae is lifted by gently inserting one of the scissor blades under the gills, and a small portion of the tissue should be excised for examination. The gill clipping from the scissor blade should be gently transferred onto the edge of a cover slip that is then positioned onto a drop of water centered on a glass slide.

Fig. 69 Removal of zebrafish operculum.

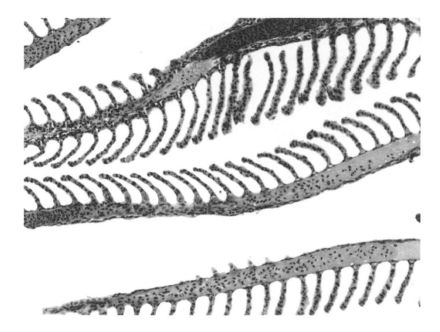

Fig. 70 Gill histomicrograph from adult zebrafish. H&E 20X. Photo courtesy of Dr. Susan Westmoreland, Assistant Professor of Pathology, Harvard Medical School, Chief, Section of Comparative Pathology.

The tissue should then be gently compressed between the cover slip and the glass slide for examination. An optimal gill sample should look like an open fan without superimposing gill lamella. Normal gill tissue should be uniformly red with well-defined lamellar margins. Secondary lamellae are usually 1 or 2 cell layers thick (**Figure 70**). Gill samples that are brown, cherry red, or pale may indicate the presence of toxicity or anemia.

Skin mucus wet mount preparation

In order to collect a mucus sample, the fish should be placed in lateral recumbency in a moist petri dish. If the fish is positioned on a dry paper towel, the mucus on the resting side will peel off when the fish is moved, resulting in a loss of diagnostic material. A mucus sample can be obtained by gently scraping the surface of the fish using a cover slip along the lateral side of the fish, moving in a cranial to caudal direction.

The mucus should accumulate on the border of the cover slip. If the sample is too small, the other side of the fish can be sampled using the same cover slip. The cover slip with the mucus sample can

Fig. 71 Two superimposed zebrafish scales.

then be positioned onto a small drop of water centered on a glass slide. Commonly identified structures include scales, mucus, and potentially ectoparasites and fungi **(Figures 71 and 72)**.

Fin wet mount preparation

The fins are extended and examined under the dissecting microscope for the presence of parasites, fungi, or any lesions. A fin biopsy can be collected using a small pair of scissors. Any fin can be biopsied. The small sample should be transferred from the scissor blade and placed on a slide with a water drop and cover slip. It is important that the fin sample lay flat on the slide to prevent the biopsy from folding onto itself.

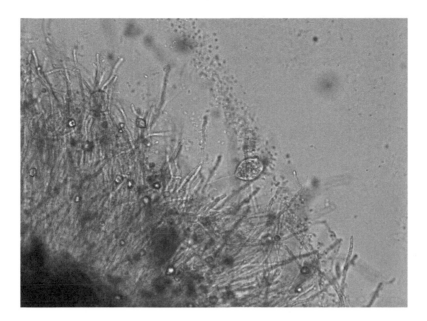

Fig. 72 Skin mucus wet mount preparation. Water mold hyphae are located in the bottom left side of the slide, and one oval ectoparasite (*Epistylis* sp.) is seen embedding within the hyphae.

Ulcerated, frayed, and inflamed tissue can be visualized microscopically. The appearance of these types of lesions is often associated with substandard water quality, parasitism, or bacterial infection.

Necropsy

Gross necropsies can be performed under a dissecting microscope with the fish in lateral recumbency in a petri dish in order to evaluate the coelomic cavity. The liver, gall bladder, spleen, swim bladder, gastrointestinal tract, and gonads are evaluated for their shape, color, and texture (**Figures 73 and 74**). A gastrointestinal tract squash preparation is commonly performed to evaluate the presence of parasites such as *Capillaria* (Kent et al. 2002).

common diseases of the laboratory zebrafish

Laboratory zebrafish are commonly housed in recirculating-water housing systems maintained in dedicated holding rooms. Pathogens

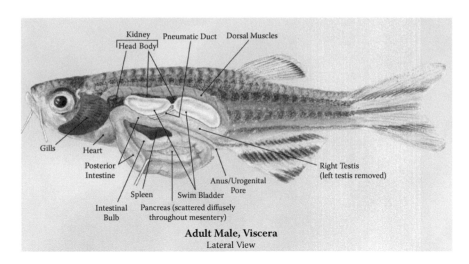

Fig. 73 Male adult zebrafish coelomic cavity. From Zebrafish Anatomy: *Danio rerio* poster. Copyright 2007, AALAS. Reprinted with permission of AALAS, Inc.

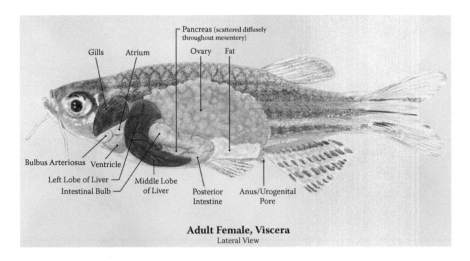

Fig. 74 Female adult zebrafish coelomic cavity. From Zebrafish Anatomy: *Danio rerio* poster. Copyright 2007, AALAS. Reprinted with permission of AALAS, Inc.

can be introduced by importing fish from sources with undefined health status, cross-contamination between aquatic species in the same institution, and poor biosecurity. Fish health or the microbial status of a population is defined by a health-screening program, which is also referred to as a "sentinel program."

In general, fish diseases have either an infectious or a non-infectious etiology. Infections are caused by organisms in the environment, including food, biofilms, intermediate hosts, and other fish. Infectious agents include parasites, bacteria, viruses, and fungi. In contrast, non-infectious diseases are caused by environmental problems, nutritional deficiencies, genetic anomalies, and trauma.

Clinical onset can be peracute, acute, or chronic. In most instances, acute onsets are due to abnormal water parameters such as low dissolved oxygen, chlorine toxicity, or supersaturation leading to rapid mortality. A low mortality rate that progresses over several days is typical of a microbial agent. Chronic illness may be characterized by lethargy, decreased appetite, reduced reproductive performance, and low mortality over extended periods of time, ranging from weeks to months. Such illness may be due to parasites, bacteria, viruses, toxins, or nutritional deficiency.

There is great variation between the responses of individual fish to pathogens, and consequently it is possible that not all fish display signs of disease during an outbreak. This variation can be influenced by genetics, environment, nutrition, and handling (Camp et al. 2000; Ortuno et al. 2003; Dror et al. 2006; Hosoya et al. 2007; Choi et al. 2007; Rodríguez et al. 2009).

Bacterial Agents and Disease

Bacterial gill disease

Flavobacterium columnare and *F. branchiophilum* are causative agents of bacterial gill disease (BGD) in freshwater fish (Kent et al. 2007). *Flavobacterium branchiophilum* is a Gram-negative filamentous bacterial rod (Schachte 1983; Noga 1996). The clinical presentation may include flared opercula, lethargy, and anorexia. A diagnosis may be made by analyzing the gills via wet mount, by histology, or by culture. Wet mount preparation of gills will reveal long, rod-shaped Gram-negative bacteria. *Flavobacterium branchiophilum* is non-motile. *Flavobacterium columnare* is a long, thin, motile bacterial rod, and tends to clump in a haystack formation. Histopathology of the gill may demonstrate proliferation of the epithelium resulting in clubbing and fusion of the lamella (Ostland et al. 1990).

Flavobacterium columnare infections are commonly associated with housing fish under stressful conditions such as overcrowding, injuries secondary to netting or handling, and subjecting fish to stressors when the fish are housed in elevated water temperatures. Incidentally, zebrafish are typically housed in water at approximately 28°C, the optimal temperature for *Flavobacterium* spp. Any condition that irritates the gill epithelium and results in inflammation, hyperplasia, or clubbing may be a predisposing factor for infection.

Flavobacterium columnare

Flavobacterium columnare was formally known as *Flexibacter columnaris, Bacillus columnaris,* and *Cytophaga columnaris. Flavobacterium columnare* is a common Gram-negative bacterium that affects the gills, skin, and fins of fish, and macroscopically it can be mistaken for a white fungus. The disease caused by *F. columnare* is also referred to as "fin rot" or "cotton wool disease" (Hoole et al. 2001). Skin damage may be necessary to initiate the infection. Clinical presentation may include white superficial lesions on the head, fins, and scales. If the gills are affected, increased respiratory rate and/or flared opercula may be observed. Skin and fin lesions may also develop into ulcers with gray-white areas, and the fins may appear ragged and frayed (Hoole et al. 2001). The lesions can be cultured for microbial isolation in order to confirm the diagnosis. Wet mounts of the lesions may show the presence of long, slender bacterial rods in haystack formation.

Aeromonas hydrophila

Aeromonas hydrophila is the etiological agent responsible for causing motile aeromonad septicemia (MAS). It is an opportunistic Gram-negative motile bacterium that infects many freshwater species, including zebrafish (Pullium et al. 1999; Rodríguez et al. 2008; Wang et al. 2009). Clinical presentation includes petechial hemorrhages of the skin, fins, oral cavity, and muscle. In some cases, superficial epidermal ulcerations leading to cavitating ulcers may develop (Palumbo et al.1989; Pullium et al. 1999; Cipriano 2001). Infection can lead to septicemia, exophthalmus, or ascites. Other findings include necrosis of the spleen, liver, kidney, and heart. Acute to chronic mortality has also been described.

Zebrafish gene expression is altered by *A. hydrophila* infection (Rodríguez et al. 2008). It has been demonstrated that expression levels of TNF alpha, IL-1 beta, and IFN gamma are upregulated in the kidneys of zebrafish exposed to viable bacteria, heat-killed bacteria, or *A. hydrophila* extracellular products. Expression levels of iNOS are upregulated by *A. hydrophila* extracellular product (Rodríguez et al. 2008). Diagnosis can be achieved by culturing the lesions at room temperature (22–25°C) using Rimmler-Shotts selective medium. Histology of the skin can reveal dermatitis and myositis. PCR analysis of tissues for the presence of *A. hydrophila* 16S rRNA gene has been described as a diagnostic tool (Wang et al. 2009).

As with most bacterial infections in fish housed in recirculating systems, important preventive measures include the minimization of stress, good husbandry practices, and stable water quality. Outbreaks in zebrafish facilities have been associated with elevated nitrite levels ranging between 1 and 5 ppm (Pullium et al. 1999). Since MAS is a stress-associated disease, resolving the underlying environmental problem may be sufficient to resolve the outbreak, and antibiotics may not be necessary.

Zebrafish embryos develop externally and so are exposed to aquatic environmental pathogens, with little ability to mount an immune response (Wang 2009). How fish embryos survive exposure to pathogens is not completely understood. Researchers have demonstrated that the maternal immunization of female zebrafish with formalin-killed *A. hydrophila* leads to a maternal increase in C3 and Bf contents, as well as a corresponding rise in the resistance and tolerance of the offspring when exposed to the bacteria (Wang 2009). Zebrafish embryos are protected against *A. hydrophila* by maternally transferred immunity of the complement system operating via the alternative pathway. Alternative complement components also play a protective role in the early zebrafish embryo (Wang 2009). Furthermore, it was demonstrated that intraperitoneal injection of beta-glucan derived from *Saccharomyces cerevisiae* protects zebrafish against *A. hydrophila* infection (Rodríguez et al. 2009).

Mycobacteriosis

Mycobacteriosis is a common disease of zebrafish. Several *Mycobacterium* species are known to infect zebrafish: *M. marinum, M. abscessus, M. chelonae, M. haemophilum, M. peregrinum,* and *M. fortuitum* (Astrofsky et al. 2000; Whipps and Kent 2006; Whipps et al. 2008; Ostland et al. 2008). *Mycobacterium* spp. are ubiquitous, acid-fast bacteria found in the soil and water.

Predisposing factors leading to infection include poor water quality, overcrowding, and immunosuppression. Affected fish must be removed from the system, since they shed bacteria from skin lesions as well as from their intestinal tracts (Whipps et al. 2008). Other modes of transmission may include ingestion of infected material, including fish tissue, food, and debris in the water. *Mycobacterium* spp. colonize biofilms on tanks and pipes, which act as a reservoir for infection (Mainous and Smith 2005; Whipps et al. 2007; Mohammad et al. 2007). Therefore, good husbandry and sanitization programs are critical for minimizing these potential sources of the infection (Whipps et al. 2007; Mohammad et al. 2007).

Mycobacteriosis in zebrafish can present as a low-grade chronic systemic bacterial infection with low mortality levels, non-healing skin or corneal ulcers, decreased reproductive performance, weight loss, and formations of granulomas (**Figure 75**) (Whipps et al. 2008). Experimentally, *M. marinum* leads to significant disease and death, while clinical presentation is less severe with *M. abscessus, M. chelonae,* and *M. peregrinum* infections (Whipps et al. 2007; Whipps et al. 2008).

The most commonly used diagnostic modalities include culture, PCR, and histology. There are various agars available to isolate *Mycobacterium* spp., such as Lowenstein-Jensen, Middlebrook LH10 agar, and 7H11 agar with oleic-acid–albumin-dextrose catalase (Astrofsky et al. 2002; Mohammad et al. 2007; Whipps et al. 2008;). Acid-fast staining is also a classic stain for diagnosing *Mycobacteria* spp. histologically; non-commercial PCR assays are also available using whole fish, spleen, and liver (Kent et al. 2004; Whipps et al. 2007; Whipps et al. 2008).

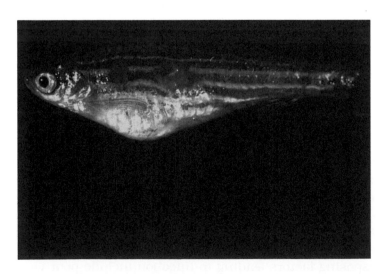

Fig. 75 Zebrafish infected with *Mycobacterium* spp. Note the distended abdomen and the skin ulcers. Photo courtesy of Dr. Trace Peterson, Department of Microbiology, Oregon State University.

The most effective disinfectants against *M. marinum* include 50% and 70% alcohol, benzyl-4-chlorophenol-2-phenylphenol (Lysol), and sodium chlorite (Clidox-S) (Mainous and Smith 2005). These agents can reduce the number of detectable *M. marinum* to zero within 1 minute of contact time (Mainous and Smith 2005). Sodium hypochlorite is moderately effective but requires at least 10 minutes of contact time to reduce bacterial counts and 20 minutes of contact time to eliminate the organism (Mainous and Smith 2005).

An important feature of *Mycobacterium* is that it is a potential zoonotic agent (Nemetz and Shotts 1993; Lehane and Rawlin 2000; Lowry and Smith 2007). *Mycobacterium marinum* is responsible for fish tank granuloma, which is also known as swimming pool granuloma (Astrofsky et al. 2002). Infection in humans potentially leads to nodules or non-healing skin ulcers, with the extremities being most commonly affected (Lowry and Smith 2007). Infection occurs when individuals are in contact with infected fish, water, and tanks while handling, cleaning, or processing fish. Individuals that are immunocompromised or who have open skin lesions are the most at risk. The best preventive measure is to wear latex gloves.

Prevention of mycobacteriosis in laboratory zebrafish entails establishing a good disinfection and quarantine program (Mainous and Smith 2005); the maintenance of favorable and stable water quality; minimizing the formation of biofilms (Mohammad et al. 2007; Whipps et al. 2007); prompt removal of sick fish; and minimization of stress (Mohammad et al. 2007; Ramsay et al. 2009; Whipps et al. 2008). There is currently no treatment available for fish.

Parasitic Agents

Pseudocapillaria tomentosa

Pseudocapillaria tomentosa is a metazoan parasite that causes intestinal capillariasis in zebrafish (**Figure 76**). It is a common nematode and has broad host specificity with the capability of infecting cyprinids and other fish (Kent et al. 2002; Moravec et al. 1999). Capillarids are pathogens known to infect all classes of vertebrates and can invade many types of tissues. Infections can negatively impact growth rates and decrease reproductive rates in fish.

Pseudocapillaria tomentosa infects the gastrointestinal tract of zebrafish, with the infection characterized by inflammation, emaciation, and intestinal carcinomas (**Figures 77 and 78**) (Kent et al.

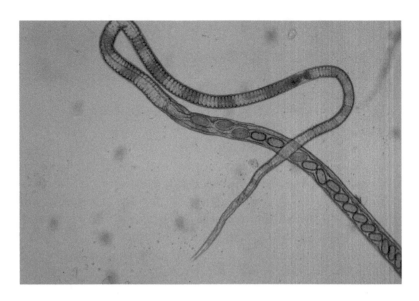

Fig. 76 Adult female *Pseudocapillaria tomentosa*. Photo courtesy of Dr. Mike Kent, Departments of Microbiology and Biomedical Sciences, Oregon State University.

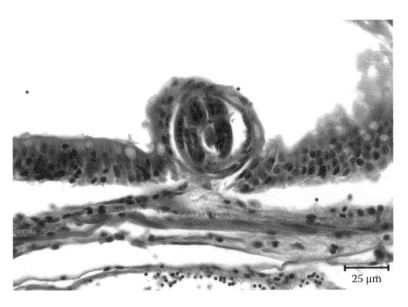

Fig. 77 *Pseudocapillaria tomentosa* burrowed into zebrafish intestinal mucosa. H&E 400X. Photo courtesy of Dr. Trace Peterson, Department of Microbiology, Oregon State University.

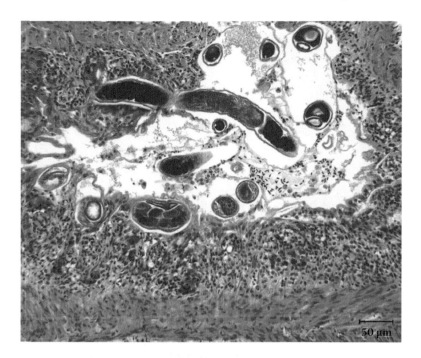

Fig. 78 *Pseudocapillaria tomentosa* causing enteritis and intestinal perforation in zebrafish. H&E 200X. Photo courtesy of Dr. Trace Peterson, Department of Microbiology, Oregon State University.

2002). *Pseudocapillaria tomentosa* can be transmitted directly and infects entire laboratory colonies (Kent et al. 2002). The life cycle of capillarids may need an intermediate host, such as the oligochaete worm *Tubifex tubifex*. Direct transmission can also occur (Kent et al. 2007).

Diagnosis can be achieved by microscopic examination of fresh fecal samples to visualize the parasite eggs and by looking at the gut content for adult worms. The worms and bioperculated eggs can also be visualized histologically using H&E staining (**Figure 79**).

Establishing a robust quarantine program is essential for preventing the introduction of *P. tomentosa* into a facility. New incoming fish should be quarantined, and feces of individual animals should be tested several times to confirm negative results. However, an even better approach is to introduce only bleached embryos into the holding facility. Other potential risks include introducing potential intermediate hosts, such as oligochaetes, into the recirculating housing systems. Currently, there are no known treatments for *Pseudocapillaria* infections in zebrafish.

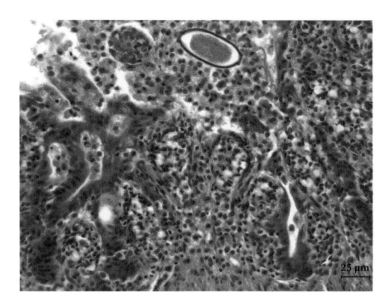

Fig. 79 *Pseudocapillaria tomentosa* bioperculated egg in zebrafish intestine lumen. H&E 400X. Photo courtesy of Dr. Trace Peterson, Department of Microbiology, Oregon State University.

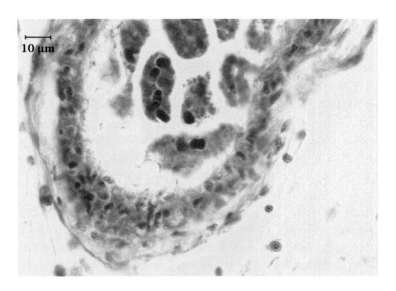

Fig. 80 Myxozoan spores in zebrafish kidney tubule. Giemsa. Photo courtesy of Dr. Mike Kent, Departments of Microbiology and Biomedical Sciences, Oregon State University.

Myxosporidiosis

Myxozoans are obligate parasites that infect aquatic vertebrates. The majority are considered non-pathogenic, but they are responsible for important economic losses in the aquaculture industry (Lom and Dykova 1992; Sitjà-Bobadilla 2008). Infections in zebrafish are addressed by the Zebrafish International Resource Center (ZIRC), where the myxozoans are described as belonging to the genera *Myxidium* and *Zschokkella* (Longshaw et al. 2005). In zebrafish they infect the common mesonephric duct and the lumens of kidney tubules and can be diagnosed by histologic examination (**Figure 80**) (Kent et al. 2007). In other fish, myxozoans can infect the skin, gills, gall bladder, and kidneys (Gratzek and Reinert 1984).

The mode of myxozoan transmission in zebrafish is unknown; however, most myxozoans require an intermediate host, usually an aquatic invertebrate such as oligochaetes, polychaetes, or bryozoans (Hallett et al. 2005; Longshaw et al. 2005; Morris and Adams 2006). As recirculating water housing systems mature, oligochaetes (*Stylaria* spp.) establish themselves as permanent residents in the troughs, pipes, tanks, and biofilters of the system. Bryozoans have also been identified in zebrafish facilities, where they may affect husbandry and potentially animal health (**Figure 81**). Bryozoans found in zebrafish facilities have been shown to harbor a variety of organisms, such as *Stylaria*, tubellarians (flatworms), and *Epistylis* (**Figure 82**). *Epistylis* is a single-celled, stalked protozoan that forms colonies and may infect the skin and fins, leading to secondary bacterial infections (**Figure 83**). Furthermore, certain oligochaetes that are sometimes used as a food source for laboratory zebrafish are also known to host parasites such as cestodes, flagellates, myxozoans, and microsporeans (Hallett et al. 2005; Kent et al. 2007).

Myxozoans that infect zebrafish target the renal system, and immunomodulation occurs as a consequence (Sitjà-Bobadilla 2008). Thus, recommending myxozoan-free animals to researchers is appropriate. Currently, treatment of zebrafish infected with myxozoa is not deemed necessary, since they are not very pathogenic and infected animals do not typically display clinical signs. However, there is so little known regarding myxozoan infection in zebrafish and its impact on health or research that it is most prudent to implement prevention programs to minimize the risk of infections.

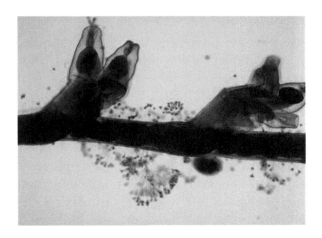

Fig. 81 Bryozoan, *Plumatella repens*, from a laboratory zebrafish research facility. The bryozoan has *Epistylis* sp. colonies attached to its surface. *Epistylis* sp., single-celled, stalked, colonial organisms with cilia around the oral opening used for feeding, can infect the skin and fins of freshwater fish. Photo courtesy of Dr. Susan Westmoreland, Assistant Professor of Pathology, Harvard Medical School, Chief, Section of Comparative Pathology.

Fig. 82 Bryozoan from a zebrafish research facility harboring *Stylaria* and tubellarian flatworms. Photo courtesy of Dr. Susan Westmoreland, Assistant Professor of Pathology, Harvard Medical School, Chief, Section of Comparative Pathology.

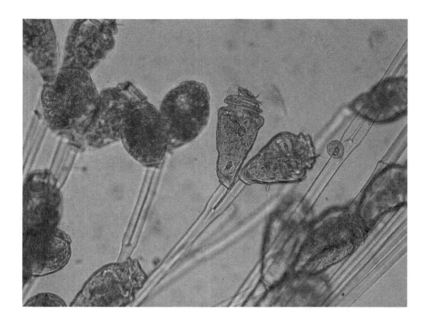

Fig. 83 *Epistylis* sp. 200X. Photo courtesy of Dr. Robert Durborow, Kentucky State University.

Metazoan Parasites

Microsporidiosis

Pseudoloma neurophilia is the etiological agent responsible for neural microsporidiosis. It is an obligate intracellular parasite and is considered one of the most common infections seen in laboratory zebrafish (Matthews et al. 2001; Kent and Bishop-Stewart 2003; Ferguson et al. 2007).

In zebrafish, *P. neurophilia* forms cysts filled with spores that can infect the brain, spinal cord, spinal ganglia, and skeletal muscle, leading to emaciation, ataxia, and spinal deformities that affect swimming patterns **(Figure 84)** (Kent et al. 2007). Microsporidian spores are resistant in the environment. Transmission occurs when zebrafish ingest them. Infection can develop 4 weeks after ingestion of infectious material, and all life stages of zebrafish are susceptible. Larval fish seem to be the most sensitive, with documented cases of high mortality in clutches assayed for survival 1 week post-exposure (Ferguson et al. 2007).

Vertical transmission has been reported in other microsporidia that infect vertebrates and invertebrates. *P. neurophilia* spores have been identified in zebrafish ovaries, as well as in immature and

Fig. 84 Zebrafish with musculoskeletal deformities and PCR positive for *Pseudoloma neurophilia.* Photo courtesy of Dr. Susan Westmoreland, Assistant Professor of Pathology, Harvard Medical School, Chief, Section of Comparative Pathology.

degenerate eggs (Kent et al. 2007). Therefore, vertical or pseudovertical (with the pathogen outside of the egg) transmission of *P. neurophilia* to progeny is possible.

Currently there are no commercially available diagnostic tests, but infection can be detected with whole-body PCR assays, wet mounts of the central nervous system, and histology demonstrating xenoma. Zebrafish that are emaciated, are ataxic, and have spinal deformities should be removed from the system to minimize the risk of spreading the disease. Good disinfection and quarantine practices can also assist in reducing the infection load. Spores are resistant to extreme environmental conditions and can survive for long periods of time. Zebrafish laboratories usually disinfect eggs to prevent transmission of pathogens, typically with chlorine at 25 to 50 ppm for 10 minutes. However, certain microsporidia spores are highly resistant to chlorine, and the commonly used egg disinfection protocols do not prevent transmission of *P. neurophilia* to progeny (Ferguson et al. 2007). Currently, there are no treatments reported to be effective against this agent in the zebrafish.

The potential impacts of *P. neurophilia* infection on research are numerous, but would mainly affect research involving musculoskeletal disease, swimming behavior, and the central nervous system.

Pleistophora hyphessobryconis

Pleistophora hyphessobryconis is a microsporidian recently discovered in laboratory zebrafish. It is the etiological agent for the neon tetra disease and is known to infect a wide range of aquarium fishes, including tetras, barbs, and goldfish (Lom and Dylova 1992). *Pleistophora hyphessobryconis* in zebrafish infects the skeletal muscle

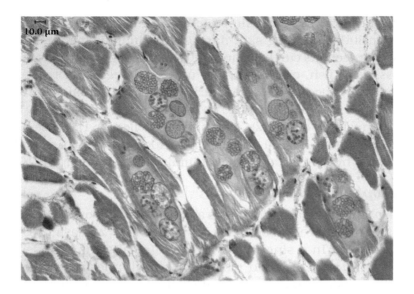

Fig. 85 *Pleistophora hyphessobryconis* infection in skeletal muscle of zebrafish. Photo courtesy of Dr. Mike Kent, Departments of Microbiology and Biomedical Sciences, Oregon State University.

and macrophages in a variety of organs (**Figure 85**). According to the ZIRC, preventing its introduction in zebrafish research facilities is recommended. Diagnosis, control, and treatment are similar to those for *Pseudoloma neurophilia*. However, a PCR test for *Pleistophora hyphessobryconis* is not yet available.

Fungal Diseases

Saprolegniasis

Saprolegniasis, caused by *Saprolegnia* spp., is the most common water mold affecting freshwater fish of all ages (Astrofsky et al. 2002). It is a ubiquitous opportunistic water inhabitant that invades traumatized epidermis in zebrafish (Astrofsky et al. 2002). Improper handling, bacterial or viral skin diseases, and trauma predispose fish to infection. Clinically, affected fish develop superficial white to brown cotton-like growths on the skin, fins, and gills. Lesions can spread rapidly and may cause death from osmotic or respiratory problems.

Saprolegniasis can be diagnosed on wet mount and culture of the skin or gills, where broad nonseptate branching hyphae that produce motile flagellated zoospores in the terminal sporangia can be visible microscopically (Noga 1996).

Water molds cannot be eliminated from any culture system; prevention includes avoiding skin damage and minimizing predisposing stressors. Infections can be acquired from fungal growth on inanimate biomass, dead fish, excess food, or unhatched eggs. Therefore, good husbandry practices that prevent accumulation of inanimate matter in the tank will minimize substrate on which the fungus can grow (Lasee 1995; Astrofsky et al. 2002). The minimum effective UV dose against *Saprolegnia* spp. is 39,600 μW sec/cm^2. *Saprolegnia* has been successfully treated by using prolonged salt immersion 1–5 ppt (Noga 1996).

Lecythophora mutabilis

Lecythophora mutabilis infection is a ubiquitous, opportunistic fungal agent that affects zebrafish fry ranging from 5 to 24 days post-hatch (Dykstra 2001). In a single case report, clinical signs included lethargy, reduced appetite, and mortality. A biofilm formed around the head of the fish occluding the oral cavity, leading to starvation and asphyxiation. Transmission between fish was direct, and the infection was diagnosed by culturing the affected fish (Dykstra 2001). In this specific case report, prevention and treatment for *L. mutabilis* infections was to maintain total hardness and calcium within normal levels (Dykstra 2001).

Parasitic Diseases

Ectoparasites found in a laboratory setting will most likely target either the skin or gills. Infected fish may have a whitish or bluish sheen on the surface of their bodies and may display an erratic swimming behavior referred to as "flashing." Increase skin mucus production due to the irritation caused by the parasite may be noted. If the gills are involved, the fish may develop respiratory distress that would be noted by increased respiratory rate, flared operculae, and lethargy. Transmission usually occurs when fish are in direct contact with infected fish or contaminated water. Infection can be diagnosed by performing either a wet mount or histology of the skin or gills. The following describes some of the ectoparasites that may be seen in laboratory zebrafish.

Chilodonella spp. is a ciliated free-swimming protozoan that targets the skin and gills. It is an oval, flat parasite with parallel rows of cilia and a notch on the anterior end of the body. Unique features of *Chilodonella* include its erratic swimming behavior and its heart-shaped appearance (Hoffman 1975). Heavy infection can lead to mortality. Fish can be treated with a salt bath.

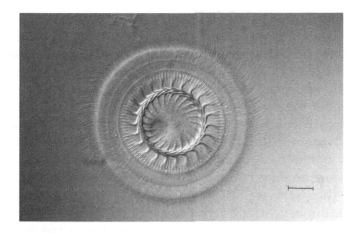

Fig. 86 Trichodinid parasite. Photo courtesy of Jayde Ferguson, Departments of Microbiology and Biomedical Sciences, Oregon State University.

Trichodina spp. is a saucer-shaped ciliated protozoan measuring 50 microns in diameter (**Figure 86**). Infection is not always associated with disease, but at a high enough parasitic load it can lead to death. Treatment may include either immersion in a formalin or salt bath.

Ichthyobodo spp. is the etiologic agent of costiosis. It is a stalked, piriform-shaped flagellate parasite measuring 6–12 microns long with flagella, and has a free swimming form (Callahan et al. 2005). Costiosis infestation may lead to mortalities. Treatment options include increasing the water temperature to 30°C or immersion in a saltwater bath.

Salt, or sodium chloride, is a mineral component of water and is commonly used to control protozoans on the gills and skin of fish. In many instances, fish can be dipped in a 3% salt solution for 30 seconds to 10 minutes, which should be effective against protozoa residing on the skin, gills, and fins of freshwater fish. After the treatment, the fish should be removed from the salt solution and transferred into clean system water.

Viruses

Endogenous retrovirus in zebrafish

Endogenous retroviruses have been identified in the genomes of most vertebrates, including humans, mice, pigs and zebrafish (Singh 2007; Wilson 2008; Kambol and Faris Abthouddin 2008).

Endogenous retroviruses are inherited in a Mendelian fashion as cellular genes, through the germline as proviral DNA. Many of them are defective, but some are intact (Singh 2007; Kambol and Faris Abthouddin 2008). They provide a large reservoir of viral genes that can be activated by factors such as mutations, carcinogens, and chemical exposures (Singh 2007). In zebrafish, two endogenous retroviruses have been discovered: ZFERV and ZFERV-2 (Shen and Steiner 2004; Kambol and Faris Abthouddin 2008). ZFERV is distinct, but phylogenetically it clusters with murine leukemia virus-related retroviruses (Shen and Steiner 2004). In mice, another very well-described endogenous retrovirus is the murine mammary tumor virus (Jacoby et al. 2002; Kambol and Faris Abtholuddin 2008). The significance of provirus in zebrafish is still unclear.

Non-Infectious Conditions

Ammonia toxicity

Ammonia is the primary waste product of fish metabolism. It is also produced when organic matters such as food, unhatched artemia, and fish decay (Shih 2008). Unionized ammonia is toxic to fish, and must be removed from the water (Noga 1996). Microbes residing in the biological filter of recirculating housing systems are responsible for the conversion of ammonia to non-toxic nitrate. A number of scenarios may disrupt the function of these microbes and lead to ammonia toxicity, including the following:

1. An excessive number of fish are added to a new system in which the bacterial population is not yet large or established enough to adequately remove the ammonia from the water.
2. The biofilter nitrifying bacteria was destroyed following an antibiotic or chemical treatment.
3. Fish shipped in bags are subject to ammonia poisoning when not packaged and unloaded properly.

Clinical signs are usually non-specific and may include hyperexcitability, anorexia, decreased respiratory rate, and reduced growth (Noga 1996). Ammonia toxicity increases the susceptibility of fish to pathogens such as *Aeromonas hydrophilia* (Pullium et al. 1999).

Prevention may be achieved by managing the system's microbial population properly and by reducing the overall ammonia production through reductions in fish density and biological load and by regular water changes. When ammonia levels are too high, water changes should be done immediately.

Nitrite toxicity

Nitrite toxicity is also known as methemoglobinemia and brown blood disease. It occurs when fish are exposed to elevated nitrite levels (Noga 1996). Toxicity depends on other water parameters such as dissolved oxygen, pH, alkalinity, and chloride levels. On average, 0.5 ppm is considered toxic to most fish (Wedemeyer 1996). The blood develops a brown-tinged appearance after nitrite oxidizes the Fe^{2+} in hemoglobin to Fe^{3+}, forming methemoglobin, which is unable to carry oxygen (Wedemeyer 1996). Similar to ammonia toxicity, nitrite toxicity is associated with a water quality problem where the microbial ecosystem harboring the nitrifying bacteria is disrupted.

The severity of the clinical presentation depends on the percentage of hemoglobin that is oxidized. Normally, fish may have 1–3% oxidized hemoglobin, and fish with levels greater than 50% will experience severe respiratory problems that may lead to death (Wedemeyer 1996). Chronic exposure to nitrite leads to gill hyperplasia, hypertrophy, lamellar separation, necrosis, and hemorrhage (Wedemeyer 1996). Methemoglinemia may include behavioral abnormalities, hypoxia, and death.

Nitrite toxicity increases with decreasing dissolved oxygen, pH, alkalinity, and chloride levels. Possible treatment of nitrite toxicity includes addition of sodium chloride or calcium chloride to the water in order to prevent methemoglobin formation (Wedemeyer 1996).

Chlorine/chloramine toxicity

Chlorine and chloramine are both highly toxic to aquatic animals (Noga 1996). Many municipalities typically add various concentrations of chlorine or chloramine to disinfect the water. Municipal water uses approximately 0.2 mg/L of total chlorine, and amounts may range from 0.5 to 1.0 mg/L (Noga 1996). Chlorine is removed by aerating the water for 24 hours and by using chemical neutralizers such as sodium thiosulfate at a concentration of at least 7.4 ppm per ppm chlorine to be neutralized (Jensen 1989; Wedemeyer 1996). However, chloramines are not removed by either aeration or neutralizers. Using a carbon filter will remove both chloramines and chlorine. Zebrafish facilities that use chlorine as a disinfectant or use municipal water without a dechlorination system are at risk.

Chlorine is also frequently used to disinfect tanks and equipment, and fish may be exposed through residues on the tanks. Clinical presentation associated with chlorine toxicity includes gill necrosis, leading to respiratory distress and asphyxiation (Noga 1996). Typically the gills are colored cherry-red in cases of chlorine toxicity.

Gas bubble disease

Gas bubble disease occurs as a result of supersaturation of the water by nitrogen gas or oxygen (Kent et al. 2007). In recirculating housing systems, this can occur under a variety of circumstances, most typically via a leaky pipe on the suction side of the pump that leads to air injection. Such a scenario leads to supersaturation of the water, where the dissolved gas exceeds the stable maximum dissolved gas levels for the local temperature, salinity, and pressure conditions.

Fish with gas bubble disease develop macroscopic bubbles in the eyes, oral mucous membranes, gills, and fins (**Figure 87**). They may also present with exophthalmia (Noga 1996). Death occurs from asphyxiation following ischemic necrosis of gill lamellae, since the air bubbles disrupt the blood flow.

Fig. 87 Zebrafish with gas bubble disease. Note air bubbles trapped in the tissue surrounding the eye. Photo courtesy of Zebrafish International Resource Center. Supported by grant P40 RR012546 from the NIH-NCRR.

preventive medicine

Preventing and managing disease outbreaks requires a comprehensive approach that includes nutritional management, water quality monitoring, quarantine programs, and sanitization practices. The pathogen–host relationship in laboratory zebrafish is a balance between the environment, the fish, and the microbial ecosystem of the recirculating water system. Disease results from a loss of equilibrium favoring the microbial ecosystem while weakening the host.

Pathogens are distinguished as being primary, secondary, or opportunistic. Primary pathogens lead to disease when water quality and the housing conditions are optimal. Secondary or opportunistic organisms lead to illness when metabolic or environmental factors are not optimal or when the animals are exposed to prolonged stress. Poor zebrafish health is often an indication that one or more environmental parameters, husbandry conditions, or experimental techniques are not adequate. Examples of environmental parameters most relevant to zebrafish include temperature, pH, hardness, alkalinity, conductivity, nutrition, high bacterial counts, excessive biofilm, injurious handling, and an inadequate post-experimental recovery period or process. Disease usually results when stressors tip the balance in favor of the pathogen (Noga 1996; Harper and Wolf 2009). Fish disease susceptibility varies with the genetics, husbandry, handling, biosecurity, and diet. Although fish do not show clinical signs, physiological changes occur as a consequence of stress and illness, impacting homeostasis, reproductive ability, growth rate, and research value.

It is well known that minimizing stress in fish is important in preventing disease outbreaks and maintaining homeostasis (Gratzek and Reinert 1984; Contreras-Sánchez et al. 1998; Barton 2002; Olsen et al. 2002; Francis-Floyd 2002; Turner et al. 2003; Dutta et al. 2005; Dror et al. 2006; Harper 2006; Pinto et al. 2007; Hosoya et al. 2007; Harper and Wolf 2009; Ramsay et al. 2009; Schreck 2009). Fish exposed to acute and chronic stress show increased susceptibility to clinical disease and morbidity (Ramsay et al. 2009). In zebrafish, stress levels can be measured by whole body cortisol levels, which increase significantly after certain husbandry practices (Ramsay et al. 2009).

Zebrafish Diets

Laboratory animal diets for research purposes should be balanced and pathogen and toxin free, since these are critical for the animals' health and research standardization (Lipman and Perkins 2002; Siccardi et al. 2009). As of this writing, standardized nutritional requirements for the life stages of zebrafish have not been determined. However, as described in Chapter 2, research is being performed on laboratory-prepared diets in order to establish an open-formulation nutritional standard to assure that nutritional variation does not compromise animal health or contribute to non-protocol-related variability in zebrafish research (Siccardi et al. 2009).

Most laboratory zebrafish are given live feed to complement their diet. This practice presents the risk of introducing environmental contaminants and pathogens into the system. For example, live food commonly given to zebrafish includes *Artemia*, worms, and *Paramecium*. Oligochaete worms are hosts to parasites, including cestodes, flagellates, microsporeans, and myxozoans (Raftos and Cooper 1990; Hallett et al. 2005; Kent et al. 2007). Myxozoan infections in zebrafish are very common. Once a myxozoan is introduced into a housing system via an infected feed, the myxozoan could theoretically complete its life cycle there, since zebrafish housing systems commonly harbor several non-pathogenic aquatic invertebrates capable of serving as an intermediate host for it, such as oligochaetes (*Stylaria* spp.), polychaetes and bryozoans (**Figure 82**) (Hallett et al. 2005; Longshaw et al. 2005; Morris and Adams 2006).

Feeding *Artemia* poses another biosecurity risk since potentially pathogenic species of bacteria are associated with *Artemia* cultures, and the abundance of certain microbes increases when the *Artemia* are enriched (Hoj et al. 2009). Therefore, the use of *Artemia* as a feed poses the threat of inadvertent introduction of potentially pathogenic organisms into housing systems.

Live feeds may also contain chemical contaminants. For instance, PCB, DDT, and heavy metals are known to accumulate in *Artemia* cysts (Olney et al. 1980; Bengtson et al. 1984; Wang et al. 2009). Various levels of chlorinated hydrocarbons and heavy metals have been identified in several strains of newly hatched *Artemia* (Olney et al. 1980). Studies have shown that *Artemia* obtained from different geographical sources will have very different LC50 results in toxicity tests (Bengtson et al. 1984).

Therefore, assuring that the food source is well balanced and devoid of pathogens and toxicants is not only critical to animal health but also to the integrity of the research. In order to be able to minimize non-protocol-related research variables, zebrafish nutrition and feeding practices need to move away from non-standardized application of live invertebrate feed toward a pathogen-free, toxin-free, and nutritionally balanced and defined formulated diet. However, such diets are only in the early stages of development (Siccardi et al. 2009). At present, the best approach is to be consistent in feeding practices and to be aware of the risks and limitations involved in the use of any available diet type—live or processed.

Source of Fish

Zebrafish purchased for research are most commonly captive bred, but are occasionally pond reared. In general, fish that are pond reared are exposed to environmental microorganisms in soil, vegetation, aquatic and avian wildlife, and other fish species reared in adjacent ponds. Purchasing pond fish and introducing them into a recirculating system presents a significant biosecurity risk by potentially introducing pathogenic organisms.

In general, pond-raised fish are reared either in a continuous production system or in a system of all-in-all-out production. A continuous production system replaces the fish that are removed from the pond with a similar number of new fish. It is simply replenishing the pond after harvest. In this system, ponds are rarely drained, and significant organic matter accumulates in the bottom of the pond. An all-in-all-out system usually drains the water at the end of each production cycle. In this practice, the pond may be washed between groups, thereby minimizing the amount of debris that accumulates at the bottom of the pond.

Wild-type laboratory zebrafish are seen as resilient fish, and when infected with pathogens they do not necessarily display clinical disease. It is important to note that transgenic and mutant-strain fish may be more susceptible to disease and are commonly kept on the same system as wild-type fish.

Disinfection of Equipment and Cleanliness

Equipment and personnel can introduce or spread pathogens within and between facilities. It is important to identify the individuals who enter the holding rooms since the high traffic flow in many facilities potentially exposes the fish to many variables. Individuals who work with fish should have a defined traffic pattern, especially if many aquatic species are maintained within the institution. For example, quarantine rooms should be visited last, and if amphibians are also being housed, fish rooms should not be entered after working with frogs.

Footbaths are common in the aquaculture industry since they may be used to decrease the overall bacterial load in the environment and prevent the introduction of pathogens into holding rooms. However, they are less frequently used in zebrafish facilities. When they are employed, footbaths are typically placed at the entrance of the fish room and between animal holding rooms.

All equipment that is used to clean tanks, such as siphon hoses, brushes, and nets, should be disinfected between uses. Disinfection can be achieved by dipping the equipment in the cleaning solution followed by a freshwater rinse. Commercial disinfectants should be used according to label instructions. After disinfecting, the equipment should be thoroughly rinsed with fresh water in order to remove any chemical residue.

Although sanitization practices have not been standardized in zebrafish facilities, the following are commonly recommended in aquaculture. Available disinfectants that can be used with equipment also include the quaternary ammonium compound Roccal-D (Winthrop Laboratories, Veterinary Products Division, New York, NY 10016); and Nolvasan-S, a brand of chlorhexidine diacetate (Fort Dodge Laboratories Inc., Fort Dodge, IA 50501).

Sodium hypochlorite is often used to clean tanks, floors, and building walls, but care must be taken to assure that adjacent tanks do not have fish as fumes can kill fish in poorly ventilated facilities (Francis-Floyd 2003). Chlorine toxicity is well documented (see above), and precautions must be taken when using it as a disinfectant to prevent accidental exposures. Various disinfecting concentrations of sodium hypochlorite have been published, including 10 mg/L for 24 hours, 200 mg/L for 30 to 60 minutes, and 100 mg/L for several hours (Francis-Floyd 2003).

Many facilities design a rotational system so that each housing unit is thoroughly cleaned once a week. Cleaning includes removal of debris by siphoning, manual removal of algae from tank walls, and removal of excess particulate matter from the biofilter or filter (Francis-Floyd 2003). Systems should be disinfected between groups of fish when possible.

In summary, maintaining a balance between the environment, the fish, and the microbial ecosystem is critical in minimizing outbreaks and generating good research. Practices should always favor the long-term homeostasis of the host when considering the pathogen–host relationship in laboratory zebrafish. This includes reducing the overall stress load to which zebrafish may be exposed and should involve evaluating the source of the fish, nutrition, husbandry practices,

handling techniques, handling frequency, experimental manipulations, and environmental parameters. A comprehensive preventive
medicine program can include

- Establishing a biosecurity program that prevents the introduction of pathogens into the facility through the shipment of
  infected fish, providing pathogen-free feed, and using a clean
  water source
- Processes that prevent disease transmission by equipment,
  personnel, or water
- Prevention maintenance programs of life support systems
- Developing quarantine and sanitization programs for the
  hatchery and grow-out facility
- Proper feed source, storage, and sanitization of food preparation area; provision of a pathogen-free source diet and/or
  minimizing, to the extent possible, the use of live feeds that
  may transmit pathogens to fish populations
- Preventing the introduction and establishment of inciden-
  tal aquatic organisms, such as oligochaetes and bryozoans,
  that may act as intermediate hosts for myxozoans and other
  pathogens

health surveillance and monitoring

Health surveillance programs are designed to evaluate fish populations
in order to prevent, detect, and manage the presence of pathogens and
the variability in husbandry practices and environmental parameters.
The objectives of these programs are to assess and mitigate contamination risks, ensure effective pathogen control, and establish biological baselines. It is well established that disease and chronic low-grade
environmental stressors can affect gene expression and cause morphological and physiological changes in fish (Olsen et al. 2002; Randall
and Tsui 2002; Dutta et al. 2005; Rodríguez et al. 2008; Siccardi et
al. 2009). Currently, the lack of standardization in zebrafish research
results in tremendous variability in all activities involved in zebrafish
research laboratories. It can thus be assumed that there will be differences between the basic biological baselines of zebrafish from different laboratories owing to differences in husbandry, handling, feeding,
water, environmental parameters, fish sources, etc. Currently, there

are no other methods to identify population changes and define baselines other than through health monitoring programs.

A commonly encountered argument challenging the usefulness of sentinel programs in zebrafish facilities is the lack of treatment standards. Monitoring programs can assist in identify the presence of pathogens that can lead to disease or, while not causing visually noticeable clinical signs, can cause histological changes or target a biological system being studied. Examples of pathogens that may act in this way are *Aeromonas hydrophila* and *Pseudoloma neurophilia*.

Currently, zebrafish health surveillance programs are not commonly used. However, the authors recommend establishing at least one sentinel tank in each system that is exposed to untreated effluent water prior to filtration. Depending on circumstances, the sentinel fish should be sent out for histopathology at least two to four times per year. The list of pathogens to be screened for may also be tailored to the specific applications of the research being performed in a facility.

anesthesia and euthanasia

The U.S. Public Health Service does not provide specific guidance on the care of fish, but it does provide guidance on the use of live embryonated eggs from avian species and other egg-laying vertebrates that develop backbones prior to hatching. These guidelines, which were developed for avian species, have been applied to the use of zebrafish embryos. In 2009, the National Institutes of Health submitted the "Final Report to OLAW on Euthanasia of Zebrafish," representing the first set of guidelines issued by a U.S. regulatory agency specific to zebrafish (NIH 2009).

Chemical Agents

Finquel® (Argent Chemical Laboratories, Inc., Redmond, WA) is the most commonly used anesthetic in zebrafish; it is commercially available as tricaine methane sulfonate (TMS). Finquel is also known as MS-222, MESAB, 3-amino benzoic acid ethyl ester, and ethyl 3-aminobenzoate (Brown 1993; Iwama and Ackerman 1994; Westerfield 2007).

MS-222 is the only anesthetic approved by the Food and Drug Administration (FDA) for use in fish. In solution, MS-222 is acidic and causes acid stress in fish (Welker et al. 2007). Addition of MS-222 to non-buffered solutions can decrease the water pH to 5, leading to acidosis in the fish (Iwama and Ackerman 1994; Noga 1996; Houston 1990). Distilled water and reverse-osmosis water are suboptimal when reconstituting MS-222 since they have very little buffering capacity. Sodium bicarbonate and sodium hydroxide are both commonly used as buffering agents (Noga 1996; Rombough 2007). Sodium bicarbonate should be added to the solution at a ratio of 2:1 (sodium bicarbonate (wt):MS-222 (wt)) and adjusted to a pH of 6 to 7 (Noga 1996). Fish are immersed in buffered water with dissolved MS-222.

MS-222 stock solution is commonly frozen in small aliquots, although the stability of frozen MS-222 is not known. MS-222 is light sensitive, and the stock solution should be stored in an opaque container and then frozen. If exposed to light, it may develop a brown tinge; if this is noted, it should be discarded (Iwama and Ackerman 1994; Noga 1996). However, the frozen stock solution should not be buffered with sodium bicarbonate, as this can cause dissociation of the sulfonate group (Houston 1990; Noga 1996). If the solution is not frozen, it should be stored properly and used within 1 month of the date of makeup.

For non-survival procedures (such as euthanasia), zebrafish ≥8 dpf should be immersed in 200–300 mg of MS-222 in 1 liter of water (NIH 2009). Necropsy can be initiated 5–10 minutes after cessation of opercular movement. Another protocol recommended for zebrafish ≥8 dpf is to anesthetize the fish with 168 mg of MS-222 in 1 liter of water followed by rapid freezing in liquid nitrogen (NIH 2009).

Cooling

Current guidelines from the AVMA state that cooling is not an acceptable method of euthanasia (AVMA 2007). However, false assumptions of ice crystal formation in the skin and in tissues of fish during cooling were refuted by recent work indicating that cooling zebrafish of a certain age is an acceptable method of euthanasia (Wilson 2009; NIH 2009). According to the *Guidelines for Use of Zebrafish in the NIH Intramural Research Program*, zebrafish ≥8 dpf can be immobilized by submersion in ice water (5 parts ice/1 part water, 0–4°C) for at least 10 minutes following cessation of opercular movement (NIH 2009). In any fish where it is difficult to visualize opercular movement, fish should be left in the ice water for at least 20 minutes after cessation of all movement to ensure death by hypoxia. For zebrafish 4–7 dpf, immobilization by submersion in ice water (5 parts ice/1 part water, 0–4°C) for at least 20 minutes can be used to ensure death by hypoxia (NIH 2009).

Bleach

Current guidelines from the AVMA do not recognize bleach as an acceptable method of euthanasia (AVMA 2007). However, according to the *Guidelines for Use of Zebrafish in the NIH Intramural Research Program*, zebrafish embryos ≤3 dpf, can be euthanized using bleach (NIH 2009). A bleach solution (sodium hypochlorite 6.15%) is added to the culture system water at 1 part bleach to 5 parts water (NIH 2009). The animals should remain in this solution at least 5 minutes prior to disposal to ensure death. As detailed in the "Scientific Background" section of the *Guidelines for Use of Zebrafish in the NIH Intramural Research Program*, pain perception has not developed at these earlier stages, so this is not considered to be a painful procedure (NIH 2009).

It is important to note that the 2009 NIH guidelines and the 2007 AVMA euthanasia guidelines are not aligned. The AVMA guidelines, while not focusing on zebrafish, do review euthanasia methods suitable for freshwater and saltwater fish raised for aquaculture, the pet fish industry, and research (AVMA 2007). Although not described specifically for zebrafish, these AVMA procedures have been shown effective for euthanizing zebrafish.

6

experimental methodology

---

restraint

Handling

Most research involving the use of zebrafish requires frequent handling of the animals. During their lifetime in a facility, fish are routinely netted and transferred from one tank to another, and certain commonly employed experimental manipulations, such as strip spawning for in vitro fertilization or fin clipping for genotyping, require direct manual manipulation. Although most laboratory strains of zebrafish have become domesticated to the extent that they are more tolerant of these kinds of disturbances (Spence et al. 2008), it is important to be mindful of the fact that all forms of handling are stressful for the animals. Therefore, these activities should be minimized to the greatest possible extent, and all people working with the fish should be trained how to properly handle them when the need arises.

Basic Handling Principles

Protection of the mucus layer

Zebrafish are covered by a mucus layer, also known as the "slime layer" or "slime coat," which covers their skin and acts as a barrier against infection. This mucus is secreted by specialized cells on the outer surface of the skin, aiding in osmoregulation and underwater motion (Helfman et al. 1997). Handling can damage the integrity of the mucus layer, and protective measures should be taken to minimize its disruption during handling. Protective measures include using soft nylon nets and placing fish on moist surfaces when procedures necessitate out-of-water manipulation. Soaps and lotions may also break down the mucus, so personnel should wash their hands before handling and wear gloves when working with the fish.

Minimizing light, noise, and vibration

Fish that are subjected to restraint or increased handling (removal from water for manipulation) may show heightened sensitivity to environmental stimuli, such as light, noise, and vibration (Ramsay et al. 2009). Higher-intensity procedures, such as chemical treatments, should be carried out in the dark or in dim light, if possible, and extreme or rapid changes in light intensities should always be avoided. Light color may also impact the stress response in fish (Johnson 2000; Volpato and Barreto 2001). In some fish, blue light seems to be the least stressful color (Volpato and Barreto 2001) Protecting the fish from direct light while they are restrained is recommended to reduce stress during handling (Hubbs et al. 1988; Wedemeyer 1996).

Loud noises and vibrations should also be minimized, as these stimuli may also be an additional source of stress to the animals during manipulations or treatments (Mullins et al. 1994). It is advisable to perform all procedures in a quiet environment where these factors can be modulated.

Anesthesia and sedation

Fish should be anesthetized for all procedures that involve their removal from the water beyond the simple act of netting and transferring them from tank to tank. The act of exposing a fish to air and handling is stressful (Ramsay et al. 2009), and the proper administration of an approved anesthetic agent, such as MS-222, helps reduce the trauma associated with the experience. The use of anesthesia also minimizes external damage to the animal since physical restraint in fish that are not sedated may damage the mucous layer or lead to injury. In some cases, it may also be beneficial to give fish mild sedatives prior to and during procedures that do not involve manipulation out of the water but where some form of restraint is required. An example of this would be chemical mutagenesis, a high-intensity procedure during which the fish become extremely excitable and stressed (Detrich et al. 1999; Canadian Council on Animal Care 2005). In these cases, an anesthetic agent, usually MS-222, is given at a dose low enough to allow the fish to maintain their equilibrium in the water column, but high enough to reduce their sensory awareness and diminish the stress induced by the procedure. For adults, this dose would be in the range of 15–50 mg/L administered via immersion.

Other general considerations

The recommended amount of time that fish should be kept out of water ranges between 30 and 90 seconds, depending on the species and procedure being performed (Ferguson and Tufts 1992; Weber and Innis 2007). The major concern is keeping the gill lamellae and skin moist. There are species-specific differences that have been identified. For example, eels and catfish have a greater ability to tolerate being out of water compared to other fish (Ferguson and Tufts 1992). As there is little published research on the air exposure tolerance of zebrafish, one should adhere to the generalized reference range described above (Ferguson and Tuftst 1992; Weber and Innis 2007). If fish must be handled repeatedly, it is advisable to give them time to recover between events (Ramsay et al. 2009). Finally, when fish are moved off-system into static holding tanks for procedures of any kind, it is critical to be mindful that water quality conditions in the static water will deteriorate over time. Fish should also be kept at low densities (fewer than 5 fish per liter) in static tanks because the rate at which water quality declines increases with fish densities, and because crowding also stresses the animals (Ramsay et al. 2006). In general, it is always good practice to work quickly when handling the fish; the sooner the animals can be returned to normal holding conditions or left undisturbed when undergoing specific procedures, the better.

Methods of Restraint

Adults and juveniles

Adult and juvenile zebrafish can be restrained manually or with a device that stabilizes the position of the anesthetized fish by maintaining dorsal or ventral recumbency. A sponge or piece of Styrofoam can be used with adult zebrafish. The position of an anesthetized fish can be stabilized by cutting a V-shaped groove in the Styrofoam or sponge using a pair of scissors or a scalpel blade **(Figure 88)** (Wooster 1993; Harms 2005). The fish is then positioned in the restrainer on its dorsum or ventrum **(Figure 89)**. The material of the restraining device should be water compatible, non-abrasive, and non-adhesive

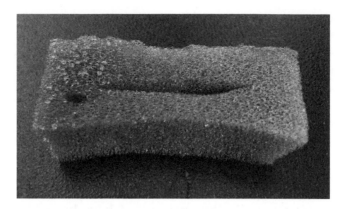

Fig. 88 Sponge with slit cut into it for securing fish.

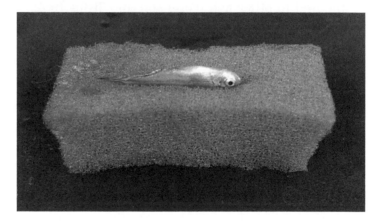

Fig. 89 Anesthetized adult zebrafish secured on sponge.

in order to protect the mucus–skin barrier. Gloves should be used when manually restraining the fish, and care should be given to the amount of pressure applied on the fish as well as to avoid direct contact with the eyes and gills.

Chamois cloths can also be used to protect the fish integument during handling. These cloths can be used while performing imaging or during transportation from the anesthetic bath to the examination table (Weber and Innis 2007).

Finquel® (Argent Chemical Laboratories, Inc. Redmond, WA), the most commonly used anesthetic in zebrafish, is commercially available as tricaine methane sulfonate (TMS). Please refer to the Anesthesia section of the chapter "Veterinary Care" for details on how to anesthetize zebrafish with MS-222.

Larval fish

The non-destructive restraint of larval zebrafish requires a different approach from that used for juvenile and adults, mostly because the animals are small and somewhat delicate at this stage. Larval zebrafish may be immobilized in a 5% methylcellulose mucilage for up to 2 weeks (King et al. 1982). This highly viscous medium prevents the fish from moving, while still allowing them to develop normally, albeit at a slightly slower rate. Upon their removal from the medium, animals are able to swim and behave in the usual fashion. This method of restraint is practical only for larval fish, largely because nutrition can be derived from the yolk and gas exchange is still taking place across the skin prior to metamorphosis.

Fasting

Fish may regurgitate under certain circumstances (Bowman 1986; Treasurer 2006). When exposed to stress, regurgitation enables the fish to conserve energy, since digestion requires energy that may be needed to fight or flee. In a research setting, fasting the fish reduces the amount of energy they use for digestion and increases the amount of energy available for recovery. In zebrafish, fasting has been used as a research tool to evaluate its impact on behavior, fin regeneration (Jowett 1999), and gene expression (Novak et al. 2005; Goldsmith et al. 2006; Amole and Unniappan 2009).

---

sampling methods

Tissue Biopsy

Collecting tissue samples in zebrafish is very common. As a genetic model, tissue samples are often taken in order to establish the presence or absence of a gene or construct (Jowett 1999). Fin samples (also referred to as fin biopsy or fin clip) are the most commonly collected tissue type, usually for the purpose of genotyping (Jowett 1999). The fish should be anesthetized when collecting fin biopsies, and precautions should be taken to protect the skin mucus and scales of the fish when handling them (Harms 2005). Anesthetized fish can be safely handled with latex gloves or when placed on a moist, non-adhesive surface. Placing fish on an adhesive surface will peel off the mucus layer from the fish as they are removed. Fin biopsies should be performed with a small, sharp pair of scissors in order to minimize tissue damage (Stoskopf 1993; Petty and Francis-Floyd 2004; Weber and Innis 2007). The sample should be small relative to the surface area of the fin. Samples can be taken longitudinally parallel to the fin rays or between the rays (Stoskopf 1993; Weber and Innis 2007).

Fin biopsies in fish are commonly performed as a diagnostic tool but can also be useful for screening for the presence of contaminants in fish (Weber and Innis 2007; Rolfhus et al. 2008; Ryba et al. 2008).

Urine

Urine collection with urethral or urinary bladder catheters has been performed to assess renal function and urinary excretion in fish such as trout, salmon, and catfish (Curtis and Wood 1991; Black 2000). Urine volumes produced by freshwater fish range from 2 to 6 mL/kg/h (Greenwell et al. 2003). Urine collection in zebrafish is not practical, however, as they do not have a bladder (Lieschke and Currie 2007).

Blood

In most fish species, physical restraint or immobilization under anesthesia is needed to collect a blood sample. Fish can be restrained physically without anesthesia, but this can cause stress and affect fish physiology and clinical chemistry (Groff and Zinkl 1999; Barton 2002; Greenwell et al. 2003; Acerete et al. 2004; Dror et al. 2006; Alsop and Vijayan 2008; Ramsay et al. 2009). The use of anesthesia during phlebotomy is highly recommended in order to minimize changes in blood parameters.

Because of their small size, blood collection in zebrafish requires anesthesia followed by euthanasia. The most common method of euthanasia in zebrafish embryos, larvae, and adults is to immerse the fish in a buffered solution of MS-222.

Blood can be sampled in zebrafish from the dorsal aorta, heart, and caudal vein (Noga 1996; Jagadeeswaran et al. 1999; Black 2000; Murtha et al. 2003). Zebrafish should be euthanized when collecting samples from the dorsal aorta. The zebrafish is positioned on a flat surface or held in one hand. With the other hand, a transverse incision is made posterior to the dorsal fin with a sharp pair of scissors. The incision should measure approximately 0.3 to 0.5 cm in depth (Jagadeeswaran et al. 1999). Pooling blood can be collected from the incision site using a micropipette tip or a capillary tube. Care should be taken not to make the incision too deep since the gastrointestinal tract could be lacerated, which would contaminate the sample (Noga 1996; Jagadeeswaran et al. 1999; Black 2000; Murtha et al. 2003).

Although in other fish the caudal vein is one of the most common sites for blood collection, the zebrafish caudal vein is difficult to access using a syringe and needle owing to its very small diameter. In small euthanized fish, the caudal vein can be accessed by cutting the base of the caudal fin with a scalpel blade followed by placing a heparinized capillary tube on the vessel (Noga 1996). However, this technique leads to tissue fluid contamination and dilution, which may affect clinical chemistry results (Ikeda and Ozaki 1981; Hogasen 1995; Congleton 2001). Although this technique has been successfully used in fish measuring 2 to 3 inches in length (Noga 1996), it is not a recommended technique for blood collection in zebrafish since it generates variable results.

Cardiocentesis, or cardiac puncture, is commonly used to collect blood from zebrafish. As with the other techniques, cardiocentesis allows for a single, terminal blood collection, because of the trauma caused to the heart (Noga 1996; Black 2000). The zebrafish heart is located in line with the posterior edge of the gills **(Figure 90)**. Zebrafish hearts are small. A one-year-old zebrafish that measures 3

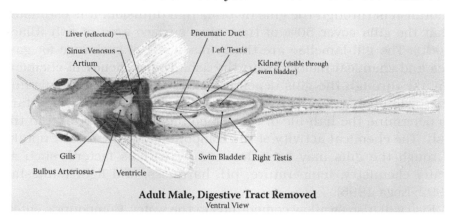

Adult Male, Digestive Tract Removed
Ventral View

Fig. 90 Illustration of an adult zebrafish demonstrating the location and anatomy of the zebrafish heart, including the bulbus arteriosus, sinus venosus, atrium, and ventricle. From Zebrafish Anatomy: *Danio rerio* poster. Copyright 2007, AALAS. Reprinted with permission of AALAS, Inc.

to 4 cm in length has a heart that can be 1 to 2 mm in length with a diameter of 1 mm (Reavill 2006).

Immediately after the fish are euthanized, blood collection may be performed by exposing the heart by making a ventral incision between the pectoral fins (Groff and Zinkl 1999). A 30-gauge needle is inserted into the ventral apex of the ventricle at a 45° angle with respect to the ventral surface of the fish. A heparinized capillary tube is then placed in the hub of the 30-gauge needle to collect the sample. If the syringe or micropipette is used, care should be taken not to hemolyse the sample by drawing air and collapsing the ventricle (Black 2000).

compound administration

Branchial Diffusion

Zebrafish are easily dosed with compounds by waterborne exposure and consequently are an attractive model for in vivo chemical exposure studies. A large number of zebrafish embryos may be rapidly evaluated in a high-throughput fashion to screen small molecules for compounds that affect development or gene expression (Murphey and Zon 2006).

One of the most common routes of administering compounds to zebrafish is through the gills by branchial diffusion. It is estimated that the gills cover 50% of the entire surface area of fish (Black 2000). The gill lamellae are the primary site of exchange for gasses and chemical molecules in the fish. The efficiency of chemical uptake through the gills depends on the size of the molecule and the hydrophobicity of the compound (Black 2000). It is important to determine the half-life of the compound in the water and in the fish. The chemical activity of the compound and the rate of uptake through the gills may be influenced by various factors, such as water chemistry, temperature, pH, hardness, and light (Lunestad 1992; Noga 1996).

Fish will also swallow compounds in the water. Compounds entering the body via this route will be absorbed through the gastrointestinal tract in addition to being absorbed through the gills.

When exposing fish to waterborne chemicals, it is important to carefully choose the compound and evaluate the potential impact of the water chemistry, chamber material, type of exposure, and vehicle on the compound's chemical activity. Some chemicals are water soluble, while others may require a vehicle such as triethylene glycol, ethanol, methanol, TWEEN 80, or dimethyl sulfoxide DMSO (Black 2000; Hallare et al. 2006).

Oral Dosing

Zebrafish may be dosed by oral gavage, or by administering compounds mixed into food or gelatin, or by bioencapsulating compounds into brine shrimp nauplii used to feed the fish (Black 2000; Schultz et al. 2007). One of the disadvantages of dosing fish with medicated feed is that the amount of food and compound consumed by the fish is difficult to quantify accurately. Depending on the compound, medicated feed can be created in the laboratory or by the fish feed manufacturer.

Anesthetized adult zebrafish can be orally dosed using a flexible rubber tube attached to a syringe or an automatic infusion pump. The tube is gently inserted into the zebrafish mouth and into the esophagus at a depth of approximately 1 cm (past the gills). The oral dose should be administered slowly to prevent perforation and regurgitation. The fish should be fasted prior to performing oral gavage in order to assure that the solution easily flows into the gastrointestinal lumen. In general, the recommended volume dose rate in fish should not exceed 1% body weight (1 mL/100 g) (Canadian Council on Animal Care 2005).

Injection

Microinjection of fish embryos is a technique widely used in zebrafish to study gene function, toxicity, and pathogenecity of microbes and to generate transgenic fish (Yuan and Sun 2009; Law 2001; Rosen 2009). It is important to note that microinjection can affect the integrity of the embryo and that training of individuals performing the microinjection is important (Tan 2008). This technique has been described in numerous publications (Black 1985; Black 2000; Xu 1999; Rosen et al. 2009).

In adult fish, the most useful injection routes are intravascular (IV), intraperitoneal (IP), and intramuscular (IM) (Noga 1996; Black 2000; Canadian Council on Animal Care 2005; Pugach et al. 2009). Compounds (or cells) can be administered IP or IM in adult zebrafish, where the needle should be positioned between the scales. IP or IM injections in adult zebrafish can be performed by using a small needle size as well as small syringes. On average, a 25-gauge needle

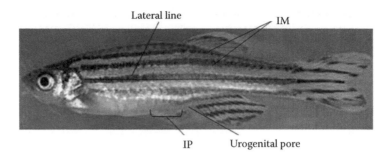

Fig. 91 Anatomical locations for intraperitoneal (IP) and intramuscular (IM) injection in adult fish. The lateral line is not visible to the naked eye; the drawn line depicts its approximate location.

is recommended, depending on the size of the fish. However, the most suitable needle gauge will depend on the viscosity of the dose vehicle.

IM injections can be administered into the dorsal epaxial and abdominal muscles; however, the lateral line and ventral blood vessels should be avoided (**Figure 91**) (Black 2000; Canadian Council on Animal Care 2005). The needle should be positioned at a 45° angle at the base of dorsal fin, caudally and ventrally onto the epaxial muscle. Compared with IP injections, the skin following an IM injection does not seal well over the injection site and the substance may leak out. Thus, to optimize the deposition of the material intramuscularly, small volumes (~0.05 mL/50 g fish) should be given slowly (Noga 1996).

When performing IP injections, the abdominal viscera should be avoided as some compounds can cause inflammation, potentially leading to the formation of adhesions (Canadian Council on Animal Care 2005). In general, the recommended fasting time for fish is approximately 24 hours prior to injections in order to minimize the risk of perforating the gut (Noga 1996). IP injections should be administered on the midline cranial to the anus (urogenital pore) and caudal to the pectoral fins (**Figure 91**) (Noga 1996). The urogenital pore must be avoided when administering compounds.

In zebrafish, administering compounds intravascularly is impractical owing to the small lumen size of the blood vessels. Although this is not a recommended approach, in certain instances zebrafish are dosed through the heart, but this leads to significant tissue damage and high mortality rates. A novel technique for administering IV injections in anesthetized zebrafish, adapted from rodent medicine, has been developed using the retro-orbital sinus (Pugach et al. 2009). Injection volumes range between 3 μL and 5 μL; compounds are administered using a Hamilton syringe (Pugach et al. 2009).

surgical procedures

Surgical Preparation

Surgery should follow the principles of the *Guide for the Care and Use of Laboratory Animals*. It is important for research personnel to be appropriately qualified and trained to ensure that good surgical technique is practiced. There are several good review articles that describe fish surgical procedures (Summerfelt and Smith 1990; Wagner et al.1999; Harms and Lewbart 2000; Harms 2005; Johnson 2000). Good surgical technique while working with zebrafish should include the use of aseptic technique; gentle tissue handling and minimal tissue dissection; the use of suitable surgical instruments; proper hemostasis management; wound closure and use of suture materials; and anesthesia and postoperative care.

Skin mucus and scales act as important barriers against infection (Harms 2005). During surgical preparation, it is important to prevent removal of the mucus and scales from the fish since this may devitalize tissue and increase the risk of secondary bacterial or saprophytic infection (Canadian Council on Animal Care 2005; Harms 2005). Swabbing the surgical site with a cotton swab with sterile saline, dilute povidone iodine, or chlorohexidine solution can help minimize superficial contamination (Harms 2005).

The skin of the zebrafish should be kept moist during the entire procedure. There are several ways to achieve this. First, the surgical site can be covered with a clear plastic sterile drape such as the 3M Steri-Drape (3M, Minneapolis, MN) (Harms 2005). A sterile drape such as this will both help prevent desiccation and protect the surgical field from contamination. Also, if a V-shaped restrainer is used, it can be soaked in water prior to surgery, thus preventing the desiccation of the scales and skin (Harms 2005).

Instrumentation

Ocular or microsurgical instruments are best suited for surgery in smaller fish such as zebrafish (Harms et al. 1995). Sterile instruments should be used to minimize contamination of the surgical site and optimize healing (Canadian Council on Animal Care 2005). Instruments should be cleaned and sterilized between surgical procedures or if inadvertently contaminated. Instruments may be gas sterilized, or autoclaved, or put in cold sterilization for 10 minutes using benzalkonium chloride or similar cold sterilants (Canadian Council on Animal Care 2005).

Wound Closure and Wound Healing

There are many factors that influence surgical wound healing in fish, including nutritional status before and after surgery; water quality; cannibalism of the surgical wound; electrolyte loss from the surgical site; and mucus layer integrity (Summerfelt and Smith 1990; Wagner et al. 1999; Johnson 2000; Harms and Lewbart 2000; Harms 2005).

Surgical sites must be closed using suture material or surgical adhesives (Canadian Council on Animal Care 2005; Harms 2005). The suture material used to close incisions in fish should be strong, inert, and non-hygroscopic, generate minimal tissue reactivity, and promote healing. Various suture types have been used in fish; the preferred ones are monofilament, such as nylon, polydioxanone, and polyglyconate (Hurty et al. 2002; Harms 2005). Monofilament material equipped with a needle with a cutting tip facilitates skin penetration (Canadian Council on Animal Care 2005; Harms 2005). Skin sutures should be removed in 10 to 14 days (Hurty et al. 2002; Harms 2005).

In fish, cyanoacrylate surgical adhesive used alone or with suture material can lead to dehiscence and delay in healing, and cause irritation. Thus the use of cyanoacrylate tissue adhesive is not recommended (Hurty et al. 2002; Harms 2005; Lowartz et al. 2005).

Anesthesia

Tricaine (MS-222) is the anesthetic agent most commonly used for zebrafish. Although there are many other anesthetics available for fish, they will not be discussed in this chapter (Iwama and Ackerman 1994; Noga 1996; AVMA 2007). The recommended stock solution of MS-222 for zebrafish is 4 g/L; the anesthetic dose is 4 mL of stock solution in 100 mL of water, or 40 µg/mL (Westerfield 2007). MS-222 is light sensitive and may develop a brown tinge if exposed to light. If this is the case, it should be discarded (Noga 1996; Iwama and Ackerman 1994).

The buffering capacity of the water has a direct effect on the impact of MS-222 on water pH. Water with no buffering capacity has zero alkalinity, and the anesthetic or euthanasia solution can lead to acidosis in fish since the water pH could decrease to 5 (Houston 1990; Iwama and Ackerman 1994; Noga 1996). Distilled water and reverse-osmosis water are suboptimal when reconstituting MS-222 since they have very little to no buffering capacity. Sodium bicarbonate or sodium hydroxide are both commonly used as buffering agents (Noga 1996). Sodium bicarbonate should be added to the solution at a ratio of 2:1 (sodium bicarbonate (wt) to MS-222 (wt)) and adjusted to a pH of 7 (Noga 1996). Fish are immersed in buffered water with dissolved MS-222.

Continuous delivery of anesthetic in water to the gills of fish is necessary for long procedures. Various delivery systems have been devised and are applicable for zebrafish (Harms and Lewbart 2000; Harms 2005; Weber and Innis 2007). However, anesthetic regimens in fish should be conducted under the advisement of a veterinarian and reviewed and approved by the IACUC.

Postoperative Care

Postoperative care in fish includes pain management and the prevention of localized infection and inflammation, peritonitis secondary to abdominal surgery, systemic infection, and loss of osmoregulation. Recovering fish should be housed separately to avoid competition for food and potential cannibalism of, and trauma to, surgical wounds. It is important to maintain clean water in the recovery tank to minimize the presence of opportunistic bacterial and fungal pathogens in order to decrease the risk of infection.

Standards for postoperative care pain management in fish have not been established. Currently there are no recommendations for postoperative analgesia in zebrafish. However, there are recommendations that have been published for fish such as koi (*Cyprinus carpio*), which include a single dose of butorphanol (Fort Dodge Animal Health, Overland Park, KS) given IM or SQ (0.4 mg/kg) prior to recovery; or a single dose of ketoprofen (Fort Dodge Animal Health) given IM at 2 mg/kg, also given prior to recovery (Harms 2005).

Research Endpoints

As with other vertebrate animal models, research endpoints should be well described in the IACUC protocol if pain and distress in fish may potentially be associated with the procedure. If no information is available regarding acceptable endpoints or expected clinical signs, a pilot study should be performed to establish the clinical signs that may be seen in order to appropriately monitor the fish on study.

Selection of appropriate endpoints is necessary to minimize the adverse effects in fish and meet the scientific goals. For example, objective evaluation of fish health status parameters or welfare indicators should be defined and described in protocols. When morbidity and mortality are expected during a study, it is necessary to define the criteria for early euthanasia. Furthermore, outlining monitoring frequency of fish that are on study will enable the removal of fish from a study prior to experiencing unnecessary morbidity.

Well-characterized descriptions of specific behavioral, physiological, and morphological parameters in healthy zebrafish will assist the research and veterinary community in defining animal welfare indicators. These indicators will be invaluable for defining research endpoints, generating reproducible data, and preventing pain and distress in fish.

resources

Because this volume is intended to be a handbook, coverage is not exhaustive for most topics. In this regard, provided here are additional resources for information related to the care and use of laboratory zebrafish.

organizations

There are a number of professional organizations that can serve as initial contacts for obtaining information regarding specific issues related to the care and use of laboratory zebrafish. Membership in these organizations should be considered, since it helps those working with zebrafish to stay abreast of various emerging regulatory, management, and health issues, as well as learn about new and improved procedures for the use of fish in research. Relevant organizations include the following:

Zebrafish Information Network (ZFIN), www.zfin.org. ZFIN is an online database that serves as the pre-eminent source of scientific information on the zebrafish. ZFIN collects, maintains, and distributes extensive information on zebrafish genetics and genomics, allowing for members of the research community to search for genes, markers, clones, antibodies, genetic maps, microarrays, and the expression patterns of selected genes. ZFIN also serves as the clearinghouse for the nomenclature of genes, transgenes, and strains of zebrafish. The information in ZFIN is constantly updated, and is extensively crosslinked to other existing scientific databases, including the National Center for Biotechnology Information (NCBI),

Ensemble, and the Zebrafish International Resource Center (ZIRC). Along with its many additional resources, the site includes a large collection of zebrafish-related publications, directories, and profiles of individuals and laboratories working with the fish, job postings, and a moderated scientific discussion forum.

Zebrafish International Resource Center (ZIRC), www.zebrafish. org. ZIRC is a sister organization to ZFIN, with a central mission to acquire resources from the research community, maintain them, and redistribute them upon request for a nominal price. The resources that ZIRC makes available for dissemination include fish strains and various genetic tools, such as expressed sequence tags (ESTs), mono-clonal antibodies, and cDNAs. ZIRC also has a health services division that provides diagnostic pathology services for zebrafish and medaka, another important model fish species. ZIRC health services has also published a manual of common diseases of zebrafish in research facilities, which is available on its website. ZIRC is the primary resource for information on the practice of zebrafish sperm cryopreservation, and researchers can submit fish from important strains to the Center for sperm collection, freezing, storage, and subsequent distribution to others in the community. ZIRC provides training and consultation on various aspects of zebrafish husbandry, both on-site in its facility and off-site at institutions that utilize the fish as a model for biomedical and genetic research.

Zebrafish Husbandry Association (ZHA), www.zhaonline.org. ZHA is an international professional organization dedicated to promoting the improvement of husbandry standards for usage of the zebrafish in scientific research. ZHA holds regular meetings, often via web-cast, and maintains a library of scientific presentations given at both ZHA-sponsored events and other events that have relevance to the zebrafish husbandry community. ZHA also convenes working groups that devise and conduct collaborative experiments in key subject areas related to the husbandry, care, and management of zebrafish. The ZHA website also includes a blog and a directory of members, among other resources. ZHA is an affiliate group of the World Aquaculture Society.

World Aquaculture Society (WAS), www.was.org. WAS is a global professional organization dedicated to the progressive and sustain-able development of aquaculture throughout the world. Membership is diverse, ranging from investigators, fish culturists, vendors, and representatives of various governmental agencies. WAS promotes aquaculture education and professional development, and publishes scientific reports on emerging topics related to the culture of aquatic plants and animals. WAS also regularly organizes and convenes a number of important scientific meetings, including the annual Aquaculture America meeting, which includes a multi-day zebrafish husbandry workshop. WAS has many affiliate organizations, including ZHA.

American Fisheries Society (AFS), www.fisheries.org. AFS is a professional organization of fisheries scientists with a mission to improve the conservation and sustainability of fishery resources and aquatic ecosystems by advancing fisheries and aquatic science and promoting the development of fisheries professionals. AFS publishes a number of scientific journals that are of interest to zebrafish professionals, including the *North American Journal of Aquaculture* and the *Journal of Aquatic Animal Health*. AFS also holds a large annual scientific meeting and maintains an excellent aquatic biology jobs board.

American Association for Laboratory Animal Science (AALAS), www.aalas.org. AALAS serves a diverse professional group, ranging from principal investigators to animal care technicians and veterinarians. The journals *Comparative Medicine* and *Journal of the American Association for Laboratory Animal Science* are both published by AALAS and serve to communicate relevant information. AALAS sponsors a program for certification of laboratory animal science professionals at three levels: assistant laboratory animal technician (ALAT), laboratory animal technician (LAT), and laboratory animal technologist (LATG). Further, a certification program for managers of animal resource programs has been developed. An extensive online resource, the AALAS Learning Library, offers subscribers an extensive menu of courses relevant to laboratory animal science. The Association also sponsors an annual meeting and several electronic listserves, including TechLink for animal technicians, CompMed for professionals working in comparative medicine and biomedical research, and IACUC-Forum for IACUC members and staff. Local groups have also organized into smaller branches.

Laboratory Animal Management Association (LAMA), www.lama-online.org. LAMA serves as a mechanism for information exchange between individuals charged with management responsibilities for laboratory animal facilities. In this regard, the association publishes the *LAMA Review* and sponsors an annual meeting.

American Society for Laboratory Animal Practitioners (ASLAP), www.aslap.org. ASLAP is an association of veterinarians engaged in laboratory animal medicine. The society publishes a newsletter to foster communication between members and sponsors sessions at the annual AALAS meeting and the annual meeting of the American Veterinary Medical Association.

American College of Laboratory Animal Medicine (ACLAM), www.aclam.org. ACLAM is an association of laboratory animal veterinarians founded to encourage education, training, and research in laboratory animal medicine. ACLAM board certifies veterinarians in the specialty of laboratory animal medicine. The group sponsors the annual ACLAM Forum continuing education meeting, along with sessions at the annual AALAS meeting.

Laboratory Animal Welfare Training Exchange (LAWTE), www.lawte.org. LAWTE is an organization of people who train in and for the laboratory animal science field. By sharing ideas on methods and materials for training, members can learn together how best to meet the training and qualification requirements of national regulations and guidelines. LAWTE holds a conference every 2 years for trainers to exchange information on its training programs in the United States and abroad.

Institute for Laboratory Animal Research (ILAR), www.dels.nas.edu/ilar. The mission of ILAR is to evaluate and disseminate information on issues related to the scientific, technological, and ethical use of animals and related biological resources in research, testing, and education. Using the principles of refinement, reduction, and replacement (3 Rs) as a foundation, ILAR promotes high-quality science through the humane care and use of animals and the implementation of alternatives. Through the reports of expert committees, the *ILAR Journal*, web-based resources, and other means of communication, ILAR functions as a component of the National Academies to provide independent, objective advice to the federal government, the international biomedical research community, and the public.

Association for Assessment and Accreditation of Laboratory Animal Care International (AAALAC International), www.aaalac.org. AAALAC is a nonprofit organization that provides a mechanism for peer evaluation of laboratory animal care and use programs.

Accreditation by AAALAC is widely accepted as strong evidence of a quality research animal care and use program.

---

publications

A number of published materials are valuable as additional reference materials, including both books and periodicals.

Books

The following books may be worthwhile sources of additional information:

1. *The Zebrafish Book*, 4th edition, by M. Westerfield. 2007. University of Oregon Press.
2. *Zebrafish: A Practical Approach*, edited by R. Dahm, C. Nusslein-Volhard, 2000. Oxford University Press.
3. *The Laboratory Fish*, edited by G. C. Ostrander, 2000. Academic Press.
4. *Plankton Culture Manual*, 6th edition, by F. Hoff and T. Snell. 2004. Florida Aqua Farms, Inc.

Periodicals

1. *Zebrafish*. Published by Mary Ann Liebert, Inc. (http://www.liebertpub.com/products/product.aspx?pid=122).
2. *Aquaculture*. Published by Elsevier (http://www.elsevier.com/wps/find/journaldescription.cws_home/503302/description#description).
3. *Aquaculture Nutrition*. Published by Wiley-Blackwell (http://www.wiley.com/bw/journal.asp?ref=1353-5773).
4. *Aquaculture International*. Published by Springer on behalf of the European Aquaculture Society (http://www.springer.com/life+sci/ecology/journal/10499).
5. *Journal of Fish Biology*. Published by Wiley-Blackwell on behalf of the Fisheries Society of the British Isles. (http://www.wiley.com/bw/journal.asp?ref=0022-1112).
6. *Journal of Fish Diseases*. Published by Wiley-Blackwell (http://www.wiley.com/bw/journal.asp?ref=0140-7775).

7. *Journal of the World Aquaculture Society.* Published by Wiley-Blackwell on behalf of the World Aquaculture Society (http://www.wiley.com/bw/journal.asp?ref=0893-8849).

8. *North American Journal of Aquaculture.* Published by the American Fisheries Society (http://afsjournals.org/loi/naja).

9. *Comparative Medicine.* Published by AALAS (http://www.aalas.org/publications/index.aspx#CM).

10. *Journal of the American Association for Laboratory Animal Science.* Published by AALAS (http://www.aalas.org/publications/index.aspx#CM).

11. *Laboratory Animals.* Published by Royal Society of Medicine Press (www.rsmjournals.com).

12. *Lab Animal.* Published by Nature Publishing Group (www.lab-animal.com).

13. *ILAR Journal.* Published by the Institute for Laboratory Animal Research (www.dels.nas.edu/ilar).

14. *ALN Magazine.* Published by Vicon Publishing, Inc. (www.animallab.com).

electronic resources

ZHA Blog. Available through ZHA. The ZHA blog can be accessed for viewing and targeted searching by anyone interested in zebrafish-husbandry-related topics, but actual posting is limited to members. Interested parties should navigate to (http://www.nezhaonline.org/post/list).

Zebrafish Newsgroup. Available through Google Groups, the Zebrafish Newsgroup is moderated by ZFIN. The newsgroup is a moderated discussion group for anyone interested in zebrafish research. To subscribe and/or view archives, interested parties should navigate to (http://groups.google.com/group/bionet.organisms.zebrafish/topics?hl=en&lnk=gschg).

CompMed. Available through AAALAS (www.aalas.org), CompMed is a listserve for the discussion of comparative medicine, laboratory animals, and topics related to biomedical research. CompMed is limited to participants who are involved in some aspect of biomedical research or veterinary medicine, including veterinarians, technicians, animal facility managers, researchers, and graduate/veterinary students. AALAS membership is not required to subscribe to CompMed.

TechLink. Also available through AALAS, TechLink is a listserve created especially for animal care technicians in the field of laboratory animal science. Open to any AALAS national member, TechLink serves as a method for laboratory animal technicians to exchange information and conduct discussions of common interest via e-mail messages with technicians in the United States and other countries around the world.

IACUC.ORG. Produced by AALAS, IACUC.ORG (www.iacuc.org) is an information resource for members and staff of institutional animal care and use committees. It is a link archive where online resources are organized by menus and submenus. IACUC.ORG was developed as an organizational tool to quickly point to a topic of interest, such as protocol forms or disaster plans used by various institutions.

AALAS Learning Library. The AALAS Learning Library provides training that is essential for technicians, veterinarians, managers, IACUC members, and investigators working with animals in a research or education setting. The courses emphasize the appropriate handling, care, and use of animals, including zebrafish.

LAWTE listserve. The Laboratory Animal Welfare Training Exchange (www.lawte.org) maintains a members-only listserve where individuals may seek information from colleagues. In addition, the site maintains a library of materials relevant to training of individuals in the proper handling and care of laboratory animals.

equipment and supplies

Please note that the following list is meant to provide examples of vendors that specialize in selling zebrafish-related equipment and supplies. This list is only partial, and other quality vendors do exist. Inclusion in this list does not imply endorsement.

Aquaneering, Inc. www.aquaneering.com. Aquaneering designs and manufactures zebrafish and *Xenopus* housing systems for research. It also supplies affiliated aquaculture equipment, feed, and system-related consumables.

Aquarienbau   Schwarz,   Inc.   www.aquaschwarz.com. Aquarienbau Schwarz designs and manufactures zebrafish and *Xenopu*s housing systems for research. It also supplies affiliated aquaculture equipment, feed, and system-related consumables.

Aquarius   Fish   Systems   www.aquariusfishsystems.com. Aquarius Fish Systems designs and manufactures zebrafish housing systems for research. It also supplies affiliated aquaculture equipment and system-related consumables.

Aquatic Eco-Systems, Inc. www.aquaticeco.com. Aquatic Eco-Systems is the primary supplier of equipment and consumables (including feeds) to the aquaculture community.

Aquatic Habitats, Inc. www.aquatichabitats.com. Aquatic Habitats designs and manufactures zebrafish and *Xenopus* housing systems for research. It also supplies affiliated aquaculture equipment, feed, and system-related consumables. It is a division of Aquatic Eco-Systems, Inc.

Brine Shrimp Direct, Inc. www.brineshrimpdirect.com. Brine Shrimp Direct is a supplier of feed and nutritional products for the aquaculture industry, including *Artemia* cysts, processed diets, and nutritional supplements. Its website also has some excellent fact sheets on *Artemia* and other feeds.

INVE Aquaculture, Inc. www.inve.com. INVE Aquaculture is a major supplier of nutritional solutions for the aquaculture industry, including processed diets, *Artemia* cysts, and nutritional supplements.

Paramecia VAP Co. www.parameciavap.com. Paramecia VAP Co. is a supplier of live paramecia cultures for the zebrafish research community.

Reed Mariculture. www.reedmariculture.com. Reed Mariculture is a major supplier of live diets and nutritional supplements for the aquaculture industry, including live rotifer cultures and Instant Algae®. Its website also has a number of excellent fact sheets on rotifers, algae, and other invertebrate food sources.

Tecniplast Aquatic Solutions www.tecniplast.it. Tecniplast Aquatic Solutions designs and manufactures zebrafish and

Xenopus housing systems for research. It also supplies affiliated equipment and system-related consumables. Tecniplast Aquatic Solutions is a division of Tecniplast, a company that designs and manufactures animal care products and related equipment.

Thoren Aquatics www.thorenaquatics.com. Thoren Aquatics designs and manufactures zebrafish housing systems for research. It also supplies affiliated aquaculture equipment and system-related consumables. Thoren Aquatics is a division of Thoren Caging Systems Inc., a designer and manufacturer of rodent caging systems and related products.

references

Acerete, L., Balasch, J. C., Espinosa, E., Josa, A., Tort, L. 2004. Physiological responses in Eurasian perch (*Perca fluviatilis, L.*) subjected to stress by transport and handling. *Aquaculture* 237:167–178.

Alsop, D., Ings, J. S., Vijayan, M. M. 2009. Adrenocorticotropic hormone suppresses gonadotropin-stimulated estradiol release from zebrafish ovarian follicle. *PLoS One* 4(7):e6463.

Alsop, D., Matsumoto, J., Brown, S., Van Der Kraak, G. 2008. Retinoid requirements in the reproduction of zebrafish. *General and Comparative Endocrinology* 156:51–62.

Alsop, D., Vijayan, M. M. 2008. Development of the corticosteroid stress axis and receptor expression in zebrafish. *American Journal of Physiology. Regulatory, Integrative and Comparative Physiology* 294:R711–719.

Alt, B., Reibe, S., Feitosa, N. M., Elsalini, O. A., Wendl T., Rohr, K. B. 2006. Analysis of origin and growth of the thyroid gland in zebrafish. *Developmental Dynamics* 235:1872–1883.

Amatruda, J. F., Patton, E. E. 2008. Genetic models of cancer in zebrafish. *International Review of Cell and Molecular Biology* 271:1–34.

Amole N., Unniappan S. 2009. Fasting induces preproghrelin mRNA expression in the brain and gut of zebrafish, *Danio rerio. General and Comparative Endocrinology* 161:133–137.

Argenton, F., Zecchin, E., Bortolussi, M. 1999. Early appearance of pancreatic hormone-expressing cells in the zebrafish embryo. *Mechanisms of Development* 87:217–221.

Arnaout, R., Ferrer, T., Huisken, J., Spitzer, K., Didier, Y. R., Stainier, M. Tristani-Firouzi, Chi, N. C. 2007. Zebrafish model for human long QT syndrome. *Proceedings of the National Academy of Sciences 104*:1316–11321.

Astrofsky, K. M., Bullis, R. A., Sagerstrom, C. G. 2002. Biology and management of the zebrafish. In *Laboratory Animal Medicine.* Amsterdam: Elsevier Science.

Astrofsky, K. M., Harper, C. M., Rogers, A. B., Fox, J. G. 2002. Diagnostic techniques for clinical investigation of laboratory zebrafish. *Lab Animal* 41–45.

Astrofsky, K. M., Schrenzel, M. D., Bullis, R. A., Smolowitz, R. M., Fox, J. G. 2000. Diagnosis and management of atypical *Mycobacterium* spp. infections in established laboratory zebrafish (*Brachydanio rerio*) facilities. *Comparative Medicine 50*:666–672.

AVMA (American Veterinary Medical Association). 2007. AVMA Guidelines on Euthanasia. *Journal of the American Veterinary Medical Association.*

Bagatto, B., Pelster, B., Burggren, W. W. 2001. Growth and metabolism of larval zebrafish: Effects of swim training. *The Journal of Experimental Biology 204*(Pt 24):4335–43.

Bailey, A. 2007. Redefining "containment" for aquatic facilities. *Animal Lab News*, January.

Bailey, A. 2008. The fish room: A machine for research. *Animal Lab News*, March.

Baker, K., Warren, K. S., Yellen, G., Fishman, M. C. 1997. Defective "pacemaker" current (Ih) in a zebrafish mutant with a slow heart rate. *Proceedings of the National Academy of Sciences 94*:4554–4559.

Barbazuk, W. B., Korf, I., Kadavi, C., Heyen, J., Tate, S., Wun, E., Bedell, J. A., McPherson, J. D., Johnson, S. L. 2000. The syntenic relationship of the zebrafish and human genomes. *Genome Research 10*:1351–8.

Barman, R. P. 1991. A taxonomic revision of the Indo-Burmese species of *Danio* Hamilton Buchanan (Pisces: Cyprinidae). *Record of the Zoological Survey of India Occasional Papers 137*:1–91.

Barton, B. A. 2002. Stress in fishes: A diversity of responses with particular reference to changes in circulating corticosteroids. *Integrative and Comparative Biology 42*:517–525.

Barut, B. A., Zon, L. I. 2000. Realizing the potential of zebrafish as a model for human disease. *Physiological Genomics* 13:49–51.

Bengtson, D. A., Beck, A. D., Lussier, S. M., Migneault D., and Olney, C. E. 1984. International Study on Artemia. XXXI. Nutritional effects in toxicity tests: Use of different *Artemia* geographical strains. In *Ecotoxicological Testing for the Marine Environment. Vol. 2*, edited by G. Persoone, E. Jaspers, and C. Claus. Ghent, Belgium.

Bennett, C. M., Kanki, J. P., Rhodes, J. 2001. Myelopoiesis in the zebrafish, *Danio rerio. Blood* 98:643 651.

Bernier, N., Bedard, N, Peter, R. E. 2004. Effects of cortisol on food intake, growth, and forebrain neuropeptide Y and corticotropin-releasing factor gene expression in goldfish. *General and Comparative Endocrinology* 135:230–240.

Bertrand, J. Y., Traver, D. 2009. Hematopoietic cell development in the zebrafish embryo. *Current Opinion in Hematology* 16:243–248.

Bird, N. C., Mabee, P. M. 2003. Developmental morphology of the axial skeleton of the zebrafish, *Danio rerio* (Ostariophysi: Cyprinidae). *Developmental Dynamics* 228:337–357.

Black, M. C. 2000. Collection of body fluids. Blood, urine, fecal materials, eggs and milt. In *The Laboratory Fish*, edited by G. K. Ostrander. Baltimore: Academic Press.

Black, M. C. 2000. Routes of administration for chemical agents. In *The Laboratory Fish*, edited by G. K. Ostrander. San Diego, CA: Academic Press.

Black, M. C., Maccubbin, A. E., Schiffert, M. 1985. A reliable, efficient, microinjection apparatus and methodology for the in vivo exposure of rainbow trout and salmon embryos to chemical carcinogens. *Journal of the National Cancer Institute* 75:1123–1128.

Boisen, A., Amstrup, J., Novak, I., Grosell, M. 2003. Sodium and chloride transport in zebrafish soft water and hard water acclimated (*Danio rerio*). *Biochimica et Biophysica Acta* 1618:207–18.

Borski, R. J, Hodson, R. G. 2003. Fish research and the Institutional Animal Care and Use Committee. *ILAR Journal* 44:286–294.

Bowman, R. E. 1986. Effect of regurgitation on stomach content data of marine fishes. *Environmental Biology of Fishes* 16:171–181.

Breder, C. M, Rosen, D. E. 1966. *Modes of Reproduction in Fishes*. New York: The Natural History Press.

Brette, F., Luxan, G., Cros, C., Dixey, H., Wilson, C., Shiels, H. A. 2008. Characterization of isolated ventricular myocytes from adult zebrafish (*Danio rerio*). *Biochemical and Biophysical Research Communications* 374:143–146.

Briggs, J. P. 2002. The zebrafish: A new model organism for integrative physiology. *The American Journal of Physiology – Regulatory, Integrative and Comparative Physiology* 282:3–9.

Brion, F., Tyler, C. R., Palazzi, X., Laillet, B., Porcher, J. M., Garric, J., Flammarion, P. 2004. Impacts of 17beta-estradiol, including environmentally relevant concentrations, on reproduction after exposure during embryo-, larval-, juvenile- and adult-life stages in zebrafish (*Danio rerio*). *Aquatic Toxicology* 68:193–217.

Brown, D. D. 1997. The role of thyroid hormone in zebrafish and axolotl development. *Proceedings of the National Academy of Sciences* 94:13011–13016.

Brown, L. A. 1993. Anaesthesia and restraint. In *Fish Medicine*, edited by M. K. Stoskopf. Philadelphia: W. B. Saunders Company.

Brownlie, A., Donovan, A., Pratt, S. J., Paw, B. H., Oates, A. C., Brugnara, C., Witkowska, H. E., Sassa, S., Zon, L. I. 1998. Positional cloning of the zebrafish sauternes gene: A model for congenital sideroblastic anaemia. *Nature Genetics* 20:244–250.

Buttner, J. K., Soderberg, R. W., Terlizzi, D. E. 1993. An introduction to water chemistry in freshwater aquaculture. Northeastern Regional Aquaculture Fact Sheet.

Cade, B. S., Terrell, J. W., Porath, M. T. 2008. Estimating fish body condition with quantile regression. *North American Journal of Fisheries Management* 28:349–359.

Cahu, C., Infante, J. Z. 2001. Substitution of live food by formulated diets in marine fish larvae. *Aquaculture* 200:161–180.

Callahan, H. A., Litaker, R. W., Noga, E. J. 2005. Genetic relationships among members of the Ichthyobodo necator complex: Implications for the management of aquaculture stocks. *Journal of Fish Diseases* 28:111–118.

Camp, K. L., Wolters, W. R., Rice, C. D. 2000. Survivability and immune responses after challenge with *Edwardsiella ictaluri* in susceptible and resistant families of channel catfish, *Ictalurus punctatus*. *Fish and Shellfish Immunology* 6:475–487.

Canadian Council on Animal Care. 2005. Guidelines on the care and use of fish in research, teaching, and testing.

Carvalho, A. P., Araujo, L., Santos, M. M. 2006. Rearing zebrafish (*Danio rerio*) larvae without live food: Evaluation of a commercial, a practical and a purified starter diet on larval performance. *Aquaculture Research* 37:1107–1111.

Chen, L. C., Martinich, R. L. 1975. Pheromonal stimulation and metabolite inhibition of ovulation in zebrafish, *Brachydanio rerio*. *Fishery Bulletin* 73:889–894.

Chen, Y. Y., Lu, F. I., Hwang, P. P. 2003. Comparisons of calcium regulation in fish larvae. *Journal of Experimental Zoology Part A–Comparative Experimental Biology* 295A:127–135.

Children's Hospital Boston. 2009. Zebrafish Genome Project.

Chizinski, C. J., Sharma, B., Pope, K. L., Pope, K. L., Patinos, R. 2008. A bioenergetic model for zebrafish *Danio rerio* (Hamilton). *Journal of Fish Biology* 73:35–43.

Choi, K., Lehmann, D. W., Harms, C. A., Law, J. M. 2007. Acute hypoxia-reperfusion triggers immunocompromise in Nile tilapia. *Journal of Aquatic Animal Health* 19:128–140.

Chu, J., Sadler, K. C. 2009. New school in liver development: Lessons from zebrafish. *Hepatology* 50:1656–63.

Cipriano, R.C. 2001. *Aeromonas hydrophila* and motile aeromonad septicemias of fish. Fish Disease Leaflet 68. Washington, D.C.: United States Department of the Interior Fish and Wildlife Service Division of Fishery Research.

Clatworthy, A. E., Lee, J. S., Leibman, M., Kostun, Z., Davidson, A. J., Hung, D. T. 2009. *Pseudomonas aeruginosa* infection of zebrafish involves both host and pathogen determinants. *Infection and Immunity* 77:1293–1303.

Cole, B., Tamuru, C. S., Bailey, R., Brown, C., Ako, H. 1999. Shipping practices in the ornamental fish industry. Center for Tropical and Subtropical Aquaculture. Publication Number 131.

Colorni, A., Diamant, A., Eldar, A., Kvitt, H., Zlotkin, A. 2002. *Streptococcus iniae* infections in Red Sea cage-cultured and wild fish. *Disease of Aquatic Organisms* 49:165–170.

Congleton, J. L., LaVoie, W. J. 2001. Comparison of blood chemistry values for samples collected from juvenile chinook salmon by three methods. *Journal of Aquatic Animal Health* 13:168–172.

Conte F. P., Wagner, H. H., Harris, T. O. 1963. Measurement of blood volume in the fish (*Salmo gairdneri gairdneri*). *American Journal of Physiology. Heart and Circulatory Physiology* 205:533–540.

Contreras-Sánchez, W. M., Schreck, C. B., Fitzpatrick, M. S., Pereira, C. B. 1998. Effects of stress on the reproductive performance of rainbow trout (*Oncorhynchus mykiss*). *Biology of Reproduction* 58:439–447.

Cortemeglia, C., Beitinger, T. L. 2005. Temperature tolerances of wild-type and red transgenic zebra danios. *Transactions of the American Fisheries Society* 134:1431–1437.

Craig, P. M., Wood, C. M., McClelland, G. B. 2007. Gill membrane remodeling with soft-water acclimation in zebrafish (*Danio rerio*). *Physiological Genomics* 30:53–60.

Craig, S., Helfrich, L. 2002. Understanding fish nutrition, feeds, and feeding. Virginia Cooperative Extension.

Crowhurst, M. O., Layton, J. E., Lieschke, G. J. 2002. Developmental biology of zebrafish myeloid cells. *International Journal of Developmental Biology* 46:483–492.

Curado, S., Stainier, D. Y., Anderson, R. M. 2008. Nitroreductase-mediated cell/tissue ablation in zebrafish: A spatially and temporally controlled ablation method with applications in developmental and regeneration studies. *Nature Protocols* 3:948–954.

Curtis, J., Wood, C. M. 1991. The function of the urinary bladder in vivo in the freshwater rainbow trout. *Journal of Experimental Biology* 155:567–583.

Dabrowski, K., Ciereszko, A. 2001. Ascorbic acid and reproduction in fish: Endocrine regulation and gamete quality. *Aquaculture Research* 32:623–638.

Dabrowski, K. R. 1986. Ontogenic aspects on nutritional requirements in fish. *Comparative Biochemistry and Physiology A–Physiology* 85:639–655.

Danilova, N., Bussmann, J., Jekosch, K., Steiner, L. A. 2005. The immunoglobulin heavy-chain locus in zebrafish: Identification and expression of a previously unknown isotype, immunoglobulin Z. *Nature Immunology* 6:295–302.

Darrow, K. O., Harris, W. A.. 2004. Characterization and development of courtship in zebrafish, *Danio rerio*. *Zebrafish* 1:40–45.

Davis, K. B., Griffin, B. R., Gray, W. L. 2002. Effect of handling stress on susceptibility of channel catfish *Ictalurus punctatus* to *Icthyophthirius multifiliis* and channel catfish virus infection. *Aquaculture* 214:55–66.

Day, A., Conlin, T., Postlethwait, J. 2009. An integrated map of the zebrafish genome (ZMAP). Zebrafish Information Network.

Deivasigamani, B. 2007. Structure of immune organ in edible catfish, *Mystus gulio. Journal of Environmental Biology 28*:757–764.

Deniz Koc, N., Aytekin, Y., Yuce, R. 2008. Ovary maturation stages and histological investigation of ovary of the zebrafish (*Danio rerio*). *Brazilian Archives of Biology and Technology 51*:513–522.

Detrich, H. W., Westerfield, M. Zon, L. I., eds. 1999. *The Zebrafish: Genetics and Genomics.* Vol. 60, *Methods in Cell Biology.* San Diego, CA: Academic Press.

Doyon, Y., J., McCammon, M., Miller, J. C., Faraji, F., Ngo, C., Katibah, G. E., Amora, R., Hocking, T. D., Zhang, L., Rebar, E. J., Gregory, P. D., Urnov, F. D., Amacher, S. L. 2008. Heritable targeted gene disruption in zebrafish using designed zinc-finger nucleases. *Nature Biotechnology 26*:702–708.

Dror, M., Sinyakov, M. S., Okun, E., Dym, M., Sredni, B., Avtalion, R. R. 2006. Experimental handling stress as infection-facilitating factor for the goldfish ulcerative disease. *Veterinary Immunology and Immunopathology 109*:279–287.

Duan, Z., Zhu, H. L., Zhu, L. Y., Yao, K., Zhu, X. S. 2008. Individual and joint toxic effects of pentachlorophenol and bisphenol A on the development of zebrafish (*Danio rerio*) embryo. *Ecotoxicology and Environmental Safety 71*:774–780.

Dutta, T., Acharya, S., Das, M. K. 2005. Impact of water quality on the stress physiology of cultured *Labeo rohita* (Hamilton-Buchanan). *Journal of Environmental Biology 26*:585–592.

Dvorak, B. I., Skipton S. O. 2008. Drinking water treatment: Reverse osmosis. University of Nebraska–Lincoln Extension, Institute of Agriculture and Natural Resources.

Dykstra, M. J., Astrofsky, K. M., Schrenzel, M. D., Fox, J. G., Bullis, R. A., Farrington, S., Sigler, L., Rinaldi, M. G., McGinnis, M. R. 2001. High mortality in a large-scale zebrafish colony (*Brachydanio rerio* Hamilton & Buchanan, 1822) associated with *Lecythophora mutabilis* (van Beyma) W. Gams & McGinnis. *Comparative Medicine 51*:361–368.

Eaton, R. C., Farley, R. D. 1974. Spawning cycle and egg-production of zebrafish, *Brachydanio rerio*, in the laboratory. *Copeia* (1):195–204.

Eaton, R. C., Farley, R. D. 1974. Growth and reduction of depensation of the zebrafish *Brachydanio rerio*, reared in the laboratory. *Copeia* 204–209.

Engeszer, R. E., Da Barbiano, L. A., Ryan, M. J., Parichy, D. M. 2005. An analysis of shoaling preference in the zebrafish, *Danio rerio. Integrative and Comparative Biology* 45:992–992.

Engeszer, R. E., Ryan, M. J., Parichy, D. M. 2004. Learned social preference in zebrafish. *Current Biology* 14:881–884.

Engeszer, R. E., Wang, G., Ryan, M. J., Parichy, D. M. 2008. Sex-specific perceptual spaces for a vertebrate basal social aggregative behavior. *Proceedings of the National Academy of Sciences of the United States of America* 105:929–933.

Engeszer, R. E., Patterson, L. B. Rao, A. A., Parichy, D. M. 2007. Zebrafish in the wild: A review of natural history and new notes from the field. *Zebrafish* 4:21–40.

Evans, D. H., Piermarini, P. M., Choe, K. P. 2005. The multifunctional fish gill: Dominant site of gas exchange, osmoregulation, acid–base regulation, and excretion of nitrogenous waste. *Physiological Reviews* 85:97–177.

Fadool, J. M., Dowling, J. E. 2008. Zebrafish: A model system for the study of eye genetics. *Progress in Retinal and Eye Research* 27:89–110.

Farber, S. A., De Rose, R. A. Olson, E. S., Halpern, M. E. 2003. The zebrafish annexin gene family. *Genome Research* 13:1082–1096.

Ferguson, H. W., Morales, J. A., Ostland, V. E. 1994. Streptococcosis in aquarium fish. *Diseases of Aquatic Organisms* 19:1–6.

Ferguson, J. A., V. Watral, A. R. Schwindt, Kent, M. L. 2007. Spores of two fish microsporidia (*Pseudoloma neurophilia* and *Glugea anomala*) are highly resistant to chlorine. *Diseases of Aquatic Organisms* 76:205–214.

Ferguson, J. A., Watral, V., Schwindt, A. R., Kent, M. L. 2007. Spores of two fish microsporidia (*Pseudoloma neurophilia* and *Glugea anomala*) are highly resistant to chlorine. *Diseases of Aquatic Organisms* 76:205–214.

Ferguson R. A., Tufts B. L. 1992. Physiological effects of brief air exposure in exhaustively exercised rainbow trout *Oncorhynchus mykiss*: Implications for "catch and release" fisheries. *Canadian Journal of Fisheries and Aquatic Sciences* 49:1157–1162.

Field, H. A., Ober, E. A., Roeser, T., Stainier, D. Y. 2003. Formation of the digestive system in zebrafish. I. Liver morphogenesis. *Developmental Biology* 253:279–290.

Fiess, J. C., Kunkel-Patterson, A., Mathias, L., Riley, L. G., Yancey, P. H., Hirano, T., Grau E. G. 2007. Effects of environmental salinity and temperature on osmoregulatory ability, organic osmolytes, and plasma hormone profiles in the Mozambique tilapia (*Oreochromis mossambicus*). *Comparative Biochemistry and Physiology Part A: Molecular and Integrative Physiology* 146:252–264.

Finney, J. L., Robertson, G. N., McGee, C. A., Smith, F. M., Croll, R. P. 2006. Structure and autonomic innervation of the swim bladder in the zebrafish (*Danio rerio*). *The Journal of Comparative Neurology* 495:587–606.

Fox, H. E., White, S. A., Kao, M. H. F., Fernald, R. D. 1997. Stress and dominance in a social fish. *Journal of Neuroscience* 17:6463–6469.

Francis-Floyd, R. 2002. Sanitation practices for aquaculture facilities. Institute of Food and Agricultural Sciences (IFAS) University of Florida, http://edis.ifas.ufl.edu/pdffiles/AE/AE08100.pdf.

Francis-Floyd, R. 2002. Stress: Its role in fish disease. Fisheries and Aquatic Sciences Department, Florida Cooperative Extension Service, Institute of Food and Agricultural Sciences, University of Florida, http://edis.ifas.ufl.edu.

Francis-Floyd, R. 2003. Sanitation practices for aquaculture facilities. Florida Cooperative Extension Service, Institute of Food and Agricultural Sciences, University of Florida. (VM87), http://edis.ifas.ufl.edu.

Gerhard, G. S., Kauffman, E. J., Wang, X., Stewart, R., Moore, J. L., Kasales, C. J., Demidenko, E., Cheng, K. C. 2002. Life spans and senescent phenotypes in two strains of zebrafish (*Danio rerio*). *Experimental Gerontology* 37:1055–1068.

Gerlach, G. 2006. Pheromonal regulation of reproductive success in female zebrafish: Female suppression and male enhancement. *Animal Behaviour* 72:1119–1124.

Gerlach, G., Lysiak, N. 2006. Kin recognition and inbreeding avoidance in zebrafish, *Danio rerio*, is based on phenotype matching. *Animal Behaviour* 71:1371–1377.

Gerlai, R., Lahav, M., Guo, S., Rosenthal, A. 2000. Drinks like a fish: Zebrafish (*Danio rerio*) as a behavior genetic model to study alcohol effects. *Pharmacology, Biochemistry and Behavior 67*:773–782.

Gerstner, J. R., Lyons, L. C., Wright, K. P. Jr, Loh, D. H., Rawashdeh, O., Eckel-Mahan, K. L., Roman, G. W. 2009. Cycling behavior and memory formation. *The Journal of Neuroscience 29*:12824–12830.

Ghysen, A., Dambly-Chaudière, C. 2004. Development of the zebrafish lateral line. *Current Opinion in Neurobiology 14*:67–73.

Goldsmith, M. I., Iovine, M. K., O'Reilly-Pol, T., Johnson, S. L. 2006. A developmental transition in growth control during zebrafish caudal fin development. *Developmental Biology 296*:450–457.

Goolish, E. M., Evans, R., Okutake, K., Max, R. 1998. Chamber volume requirements for reproduction of the zebrafish *Danio rerio*. *Progressive Fish-Culturist 60*:127–132.

Goolish, E. M., K. Okutake, S. Lesure. 1999. Growth and survivorship of larval zebrafish *Danio rerio* on processed diets. *North American Journal of Aquaculture 61*:189–198.

Gratzek, J. B., Reinert, R. 1984. Physiological responses of experimental fish to stressful conditions. *National Cancer Institute Monograph 65*:187–193.

Greenwell, M. G., Sherrill, J., Clayton, L. A. 2003. Osmoregulation in fish mechanisms and clinical implications. *The Veterinary Clinics of North America Exotic Animal Practice 6*:169–189.

Groff, J. M., Zinkl, J. G. 1999. Hematology and clinical chemistry of cyprinid fish: Common carp and goldfish. *The Veterinary Clinics of North America Exotic Animal Practice 3*:741–776.

Guo, S. 2004. Linking genes to brain, behavior and neurological diseases: What can we learn from zebrafish? *Genes, Brain and Behavior 3*:63–74.

Hallare, A., K., Nagel, H. R., Kohler, Triebskorn, R. 2006. Comparative embryotoxicity and proteotoxicity of three carrier solvents to zebrafish (*Danio rerio*) embryos. *Ecotoxicology and Environmental Safety 63*:378–388.

Hallett, S. L., Atkinson, S. D. Erséus, C., El-Matbouli, M. 2005. Dissemination of triactinomyxons (Myxozoa) via oligochaetes used as live food for aquarium fishes. *Disease of Aquatic Organisms 65*:137–152.

Hansen, A., Reutter, K., Zeiske, E. 2002. Taste bud development in the zebrafish, *Danio rerio*. *Developmental Dynamics* 223:483–496.

Harder, W. 1975. *Anatomy of Fishes: Part I. Text. Part 2. Figures and Plates*. Stuttgart: Schweizerbart.

Harms, C. A. 2005. Surgery in fish research: Common procedures and postoperative care. *Lab Animal* 34:28–34.

Harms, C. A., Lewbart, G. A. 2000. Surgery in fish. *The Veterinary Clinics of North America. Exotic Animal Practice* 3:759–774.

Harms, C. A., Bakal, R. S., Khoo, L. H., Spaulding, K. A., Lewbart, G. A. 1995. Microsurgical excision of an abdominal mass in a gourami. *Journal of the American Veterinary Medical Association* 207:1215–1217.

Harper, C., Wolf, J. C. 2009. Morphologic effects of the stress response in fish. *ILAR Journal* 50:387–397.

Harper, C. 2006. Aquatic animal stress and health management. *Aquaculture Magazine*.

Harris, J., Bird, D. J. 2000. Modulation of the fish immune system by hormones. *Veterinary Immunology and Immunopathology* 77:163–176.

Helfman, G. S., Collette, B. B. Facey, D. E. 1997. *The Diversity of Fishes*. Oxford: Blackwell Science.

Her, G. M., Chiang, C. C., Wu, J. L. 2004. Zebrafish intestinal fatty acid binding protein (I-FABP) gene promoter drives gut-specific expression in stable transgenic fish. *Genesis* 38:26–31.

Hirai, N., Nanba, A., Koshio, M., Kondo, T., Morita, M., Tatarazako, N. 2006. Feminization of Japanese medaka (*Oryzias latipes*) exposed to 17beta-estradiol: Formation of testis-ova and sex-transformation during early ontogeny. *Aquatic Toxicology* 77:78–86.

Hirata, M., Nakamura, K., Kanemaru, T., Shibata, Y., Kondo, S. 2003. Pigment cell organization in the hypodermis of zebrafish. *Developmental Dynamics* 227:497–503.

Hoffman, G. L., Landolt, M., Camper, J. E., Coats, D. W., Stookey, J. L., Burek, J. D. 1975. A disease of freshwater fishes caused by *Tetrahymena corlissi* Thompson, 1955, and a key for identification of holotrich ciliates of freshwater fishes. *The Journal of Parasitology* 61:217–223.

Hogasen, H. R. 1995. Changes in blood composition during sampling by caudal vein puncture or caudal transection of the teleost *Salvelinus alpinus. Comparative Biochemistry and Physiology Part A: Physiology 111*:99–105.

Hoj, L., Bourne, D. G., Hall, M. R. 2009. Localization, abundance and community structure of bacteria associated with *Artemia*: Effects of nauplii enrichment and antimicrobial treatment. *Aquaculture 293*:278–285.

Hoole, D. Bucke, D., Burgess, P., Wellby, I. 2001. *Diseases of Carp and Other Cyprinid Fishes*, edited by F. N. Books. Oxford [England], Malden, MA: Blackwell Science.

Hoshijima, K., Hirose, S. 2007. Expression of endocrine genes in zebrafish larvae in response to environmental salinity. *Journal of Endocrinology 193*:481–491.

Hosoya, S., Johnson, S .C., Iwama, G. K., Gamperl, A. K., Afonso, L. O. 2007. Changes in free and total plasma cortisol levels in juvenile haddock (*Melanogrammus aeglefinus*) exposed to long-term handling stress. *Comparative Biochemistry and Physiology. Part A: Molecular and Integrative Physiology 146*:78–86.

Houston, A. H. 1990. Blood and circulation. In *Methods in Fish Biology*, edited by C. B. Schreckand and P. B. Moyle. Bethesda, MD: American Fisheries Society.

Höglund, E. P., Balm, H. M., Winberg, S. 2000. Skin darkening, a potential social signal in subordinate Arctic charr (*Salvelinus alpinus*): The regulatory role of brain monoamines and proopiomelanocortin-derived peptides. *The Journal of Experimental Biology 203*:1711–1721.

Höglund, E. P., Balm, H. M., Winberg, S. 2002. Behavioural and neuroendocrine effects of environmental background colour and social interaction in Arctic charr (*Salvelinus alpinus*). *The Journal of Experimental Biology 205*:2535–2543.

Hrubec, T. C., Smith, S. A. 1999. Differences between plasma and serum samples for the evaluation of blood chemistry values in rainbow trout, channel catfish, hybrid tilapias and striped bass. *Journal of Aquatic Animal Health 11*:116–122.

Hrubec, T. C., Robertson, J. L., Smith, S. A. 1997. Effects of temperature on hematologic and biochemical profiles of sunshine bass (*Morone chrysops x Morone saxatilis*). *American Journal of Veterinary Research 58*:126–130.

Hu, N., Yost, H. J., Clark, E. B. 2001. Cardiac morphology and blood pressure in the adult zebrafish. *The Anatomical Record* 264:1–12.

Huang, H. T., Zon, L. I. 2008. Regulation of stem cells in the zebra fish hematopoietic system. *Cold Spring Harbor Symposia on Quantitative Biology* 73:111–118.

Hubbs C., Nickum J. G., Hunter J. R. 1988. Guidelines for the use of fish in research. *Fisheries* 13:16–22.

Huising, M. O., Metz, J. R., De Mazon, A. F., Lidy Verburg-van Kemenade, B. M., Flika, G. 2005. Regulation of the stress response in early vertebrates. *Annals of the New York Academy of Sciences* 1040:345–347

Hurty, C. A., Brazik, D. C., Law, J. M., Sakamoto, K., Lewbart, G. A. 2002. Evaluation of the tissue reactions in the skin and body wall of koi (*Cyprinus carpio*) to five suture materials. *The Veterinary Record* 151:324–328.

Hwang, P. P. 2009. Ion uptake and acid secretion in zebrafish (*Danio rerio*). *The Journal of Experimental Biology* 212:1745–1752.

Ikeda, Y., Ozaki, H. 1981. The examination of tail peduncle severing blood sampling method from aspect of observed serum constituent levels in carp. *Bulletin of the Japanese Society of Scientific Fisheries* 47:1447–1453.

Institute, Welcome Trust Sanger. 2009. *Danio rerio* Sequencing Project.

Iwama, G. K. 2007. The welfare of fish. *Diseases of Aquatic Organisms* 75:155–158.

Iwama, G. K., Ackerman, P. A. 1994. Anaesthesia. In *Biochemistry and Molecular Biology of Fishes*, Vol. 3. edited by P.W. Hochachka and T.P. Mommsen, Amsterdam: Elsevier Science B.V.

Iwama, G. K., Afonso, L., Todgham, A., Ackerman, P., Nakano, K. 2004. Commentary: Are hsps suitable for indicating stressed states in fish? *The Journal of Experimental Biology*, 207:15–19.

Jacoby, R. O., Fox, J. G., Davisson, M. 2002. Biology and diseases of mice. In *Laboratory Animal Medicine*, edited by J. G. Fox, L. C. Anderson, F. M. Loew, and F. W. Quimby. Orlando: Academic Press.

Jagadeeswaran, P., Liu, Y. C., Sheehan, J. P. 1999. Analysis of hemostasis in the zebrafish. *Methods in Cell Biology* 59:337–357.

Jagadeeswaran, P., Sheehan J. P., Craig, F. E., Troyer, D. 1999. Identification and characterization of zebrafish thrombocytes. *British Journal of Haematology* 107:731–738.

Jaillon, O., Aury, J. M., Brunet F., Petit, J. L., Stange-Thomann, N. E., Mauceli, L. Bouneau, et al. 2004. Genome duplication in the teleost fish *Tetraodon nigroviridis* reveals the early vertebrate proto-karyotype. *Nature 431*(7011):946–57.

Jardine, D., Litvak, M. K. 2003. Direct yolk sac volume manipulation of zebrafish embryos and the relationship between offspring size and yolk sac volume. *Journal of Fish Biology 63*:388–397.

Jaya-Ram, A., Kuah, M. K., Lim, P. S., Kolkovski, S., Shu-Chien, A. C. 2008. Influence of dietary HUFA levels on reproductive performance, tissue fatty acid profile and desaturase and elongase mRNAs expression in female zebrafish *Danio rerio. Aquaculture 277*:275–281.

Jensen, G. 1989. *Handbook for Common Calculations in Finfish Aquaculture*, Publication 8903. Baton Rouge: Louisiana State University, Agricultural Center.

Johansen, R., Needham, J. R., Colquhoun, D. J. 2006. Guidelines for health and welfare monitoring of fish used in research. *Laboratory Animals 40*:323–340.

Johnson, G. R. 2000. Surgical Techniques. In *The Laboratory Fish*, edited by G. K. Ostrander. San Diego, CA: Academic Press.

Johnson J. C., Nettikadan, S. R., Vengasandra, S. G., Lovan S., Muys, J., Henderson, E., Christiansen, J. 2005. Characterization of testudine melanomacrophage linear, membrane extension processes—cablepodia—by phase and atomic force microscopy. *In Vitro Cellular and Developmental Biology 41*:225–231.

Jonz, M. G., Nurse, C. A. 2003. Neuroepithelial cells and associated innervation of the zebrafish gill: A confocal immunofluorescence study. *The Journal of Comparative Neurology 461*:1–17.

Jonz, M. G., Nurse, C. A. 2005. Development of oxygen sensing in the gills of zebrafish. *The Journal of Experimental Biology 208*:1537–1549.

Jonz, M. G., Nurse, C. A. 2008. New developments on gill innervation: Insights from a model vertebrate. *The Journal of Experimental Biology 211*:2371–2378.

Jowett, T. 1999. Transgenic zebrafish. In *Molecular Embryology Methods and Protocols*, edited by I. Mason and P.T. Sharpe. Totowa, NJ: Humana Press.

Jørgensen, A., Morthorst, J. E., Andersen, O., Rasmussen, L. J., Bjerregaard, P. 2008. Expression profiles for six zebrafish genes during gonadal sex differentiation. *Reproductive Biology and Endocrinology* 6:25.

Kambol, R., Faris Abtholuddin, M. 2008. *Genome Structure and Characterisation of an Endogenous Retrovirus from the Zebrafish Genome Project Database*. Bioinformatics Research and Development.

Kelsch, S. W., Neill, W. H. 1990. Temperature preference versus acclimation in fishes: Selection for changing metabolic optima. *Transactions of the American Fisheries Society 119*:601–610.

Kent, M. L., Spitsbergen, J., Matthews, J. M., Fournie, J. W., Westerfield, M. 2007. *ZIRC Health Services Zebrafish Disease Manual*. Diseases of zebrafish in research facilities. Zebrafish International Resource Center.

Kent, M. L., Bishop-Stewart, J. K., Matthews, J. L., Spitsbergen, J. M. 2002. *Pseudocapillaria tomentosa*, a nematode pathogen, and associated neoplasms of zebrafish (*Danio rerio*) kept in research colonies. *Comparative Medicine 52*:354–358.

Kent, M. L., Bishop-Stewart, J. K. 2003. Transmission and tissue distribution of *Pseudoloma neurophilia* (Microsporidia) of zebrafish, *Danio rerio* (Hamilton). *Journal of Fish Diseases 26*:423–426.

Kent, M. L., Whipps, C. M., Matthews, J. L., Florio, D., Watral, V., Bishop-Stewart, J. K., Poort, M., Bermudez, L. 2004. Mycobacteriosis in zebrafish (*Danio rerio*) research facilities. *Comparative Biochemistry and Physiology. C: Comparative Pharmacology and Toxicology. 138*:383–390.

Kerr, J. P. 1962. Grouping behavior of the zebrafish as influenced by social isolation. *American Zoologist 2*:532–533.

Kim, H. J., Sumanas, S., Palencia-Desai, S., Dong, Y., Chen, J. N., Lin, S. 2006. Genetic analysis of early endocrine pancreas formation in zebrafish. *Molecular Endocrinology (Baltimore, Md.) 20*:194–203.

Kimmel, C. B., Ballard, W. W., Kimmel, S. R., Ullmann, B., Schilling, T. F. 1995. Stages of embryonic development of the zebrafish. *Developmental Dynamics 203*:253–310.

King, G. M., Gordon, R., Karmali, K., Biberman, L. J. 1982. A new method for the immobilization of teleost embryos for time-lapse studies of development. *Journal of Experimental Zoology* 220:147–151.

Kishi, S. 2006. Zebrafish as aging models. In *Handbook of Models for Human Aging*, edited by P. M. Conn. Burlington, MA: Elsevier.

Kleinjan, D. A., Bancewicz, R. M., Gautier, P., Dahm, R., Schonthaler, H. B., Damante, G., Seawright, A., Hever, A. M., Yeyati, P. L., van Heyningen, V., Coutinho, P. 2008. Subfunctionalization of duplicated zebrafish pax6 genes by cis-regulatory divergence. *PLoS Genetics* 4(2):e29.

Kobayashi, I., Ono, H., Moritomo, T., Kano, K., Nakanishi, T., Suda, T. 2010. Comparative gene expression analysis of zebrafish and mammals identifies common regulators in hematopoietic stem cells. *Blood* 115:e1–9.

Kusuda R., Salati, F. 1999. *Enterococcus seriolicida* and *Streptocuccus iniae*. In *Fish Diseases and Disorders; Viral, Bacterial and Fungal Infections*, edited by P.T.K. Woo and D.W. Bruno. Wallingford, UK: CABI Publishing.

Kuwada, H., Masuda, R., Shiozawa, S., Kogane, T., Imaizumi, K., Tsukamoto, K. 2000. Effect of fish size, handling stresses and training procedure on the swimming behavior of hatchery-reared striped jack: Implications for stock enhancement. *Aquaculture* 185:245–256.

Laale, H. W. 1977. Biology and use of zebrafish, *Brachydanio rerio*, in fisheries research – Literature review. *Journal of Fish Biology* 10:121–173.

Larson, E. T., O'Malley, D. M., Melloni, R. H. 2006. Aggression and vasotocin are associated with dominant–subordinate relationships in zebrafish. *Behavioural Brain Research* 167:94–102.

Lasee, B. 1995. *Introduction to Fish Health Management*. Omalaska, WI: U.S. Fish and Wildlife Service.

Law, J. M. 2001. Mechanistic considerations in small fish carcinogenicity testing. *ILAR Journal* 42:274–284.

Lawrence, C. 2007. The husbandry of zebrafish (*Danio rerio*): A review. *Aquaculture* 269:1–20.

Lawrence, C., Ebersole, J. P., Kesseli, R. V. 2008. Rapid growth and out-crossing promote female development in zebrafish (*Danio rerio*). *Environmental Biology of Fishes* 81:239–246.

Lehane, L., Rawlin, G. T. 2000. Topically acquired bacterial zoonoses from fish: A review. *The Medical Journal of Australia* 173:256–259.

Leung, A. Y., Mendenhall, E. M., Kwan, T. T., Liang, R., Eckfeldt, C., Chen E., Hammerschmidt, M., Grindley, S., Ekker, S. C., Verfaillie, C. M. 2005. Characterization of expanded intermediate cell mass in zebrafish chordin morphant embryos. *Developmental Biology* 277:235–254.

Lewbart, G. A. 2002. Reproductive medicine in koi (*Cyprinus carpio*). *Veterinary Clinics of North America: Exotic Animal Practice* 5:637–648.

Lieschke, G. J., Currie, P. D. 2007. Animal models of human disease: Zebrafish swim into view. *Nature Reviews Genetics* 5:353–367.

Lieschke, G. J., Oates, A. C., Crowhurst, M. O., Ward, A. C., Layton, J. E. 2001. Morphologic and functional characterization of granulocytes and macrophages in embryonic and adult zebrafish. *Blood* 98:3087–3096.

Lieschke, G., Trede, N. 2009. Fish immunology. *Current Biology* 19:R678–R682.

Lieschke, G. J., Currie, P. D. 2007. Animal models of human disease: Zebrafish swim into view. *Nature Reviews Genetics* 8:353–367.

Lieschke, G. J., Oates, A. C., Paw, B. H., Thompson, M. A., Hall, N. E., Ward, A. C., Ho, R. K., Zon, L. I., Layton, J. E. 2002. Zebrafish SPI-1 (PU.1) marks a site of myeloid development independent of primitive erythropoiesis: Implications for axial patterning. *Developmental Biology* 246:274–295.

Lim, L. C., Dhert, P., Sorgeloos, P. 2003. Recent developments and improvements in ornamental fish packaging systems for air transport. *Aquaculture Research* 34:923–935.

Lin, H. F., Traver, D., Zhu, H., Dooley, K., Paw, B. H., Zen, L. I., Handin, R. I. 2005. Analysis of thrombocyte development in CD41–GFP transgenic zebrafish. *Blood* 106:3803–3810.

Lindsay S. M., Vogt, R. G. 2004. Behavioral responses of newly hatched zebrafish (*Danio rerio*) to amino acid chemostimulants. *Chemical Senses* 29:93–100.

Lipman, N., Perkins, S. 2002. Factors that may influence animal research. In *Laboratory Animal Medicine*, edited by J. G. Fox, L. C. Anderson, F. M. Loew, and F. W. Quimby. Orlando: Academic Press.

Liu, N. A., Liu, Q., Wawrowsky, K., Yang, Z., Lin, S., Melmed, S. 2006. Prolactin receptor signaling mediates the osmotic response of embryonic zebrafish lactotrophs. *Molecular Endocrinology (Baltimore, Md.)* 20:871–880.

Liu, Y., Chan, W. 2002. Thyroid hormones are important for embryonic to larval transitory phase in zebrafish. *Differentiation* 70:36–45.

Liu, Y. W. 2007. Interrenal organogenesis in the zebrafish model. *Organogenesis* 3:44–48.

Lom, J. Dyková, I. 1992. *Protozoan Parasites of Fishes.* Vol. 26, *Developments in Aquaculture and Fisheries Science.* Amsterdam, Netherlands: Elsevier Sciences.

Longshaw, M., Frear, P. A., Feist, S. W. 2005. Descriptions, development and pathogenicity of myxozoan (Myxozoa: Myxosporea) parasites of juvenile cyprinids (Pisces: Cyprinidae). *Journal of Fish Diseases* 28:489–508.

Lowartz, S. M., Holmberg, D. L., Ferguson, H. W., Beamish, F. W. H. 2005. Healing of abdominal incisions in sea lamprey larvae: A comparison of three wound-closure techniques. *Journal of Fish Biology* 54:616–626.

Lowry, T, Smith, S. A. 2007. Aquatic zoonoses associated with food, bait, ornamental, and tropical fish. *Journal of the American Veterinary Medical Association* 231:876–880.

Lunestad, B.T. 1992. Fate and effects of antibacterial agents in aquatic environments. In *Chemotherapy in Aquaculture: From Theory to Reality*, edited by M. C. Alderman. Paris: Office International de Epizooties.

MacPhail, R. C., Brooks, J., Hunter, D. L., Padnos, B., Irons, T. D., Padilla, S. 2009. Locomotion in larval zebrafish: Influence of time of day, lighting and ethanol. *Neurotoxicology* 30:52–58.

Mainous, M., Smith, S. 2005. Efficacy of common disinfectants against *Mycobacterium marinum. Journal of Aquatic Health* 17:284–288.

Marques, I. J., Leito, J. T. D., Spaink, H. P., Testerink, J., Jaspers, R. T., Witte, F., van den Berg, S., Bagowski, C. P. 2008. Transcriptome analysis of the response to chronic constant hypoxia in zebrafish hearts. *Journal of Comparative Physiology. B: Biochemical, Systemic, and Environmental Physiology* 178:77–92.

Masser, M. P., Rakocy, J., Losordo. T. 1999. Recirculating aquaculture tank production systems: Management of recirculating systems. Southern Regional Aquaculture Publication.

Masuda, R., Tsukamoto, K., Imaizumi, K., Shiozawa, S., Sekiya, S., Nishi, A. 1993. Spiral diving behaviour and horizontal movement in juvenile striped jack *Pseudocaranx dentex* after the release in the sea. *Academic Journal* 22:49–53.

Matthews, J. L., Brown, A. M., Larison, K., Bishop-Stewart, J. K., Rogers, P., Kent, M. L. 2001. *Pseudoloma neurophilia* n.g., n.sp., a new microsporidium from the central nervous system of the zebrafish (*Danio rerio*). *The Journal of Eukaryotic Microbiology* 48:227–233.

Matthews, M., Trevarrow, B., Matthews J. 2002. A virtual tour of the Guide for zebrafish users. *Lab Animal* 31:34–40.

Matthews, R. P., Lorent, K., Russo, P., Pack, M. 2004. The zebrafish onecut gene hnf-6 functions in an evolutionarily conserved genetic pathway that regulates vertebrate biliary development. *Developmental Biology* 274:245–59.

McCarthy, I. D., Carter, C. G., Houlihan, D. F. 1992. The effect of feeding hierarchy on individual variabilty in daily feeding of rainbow trout, *Oncorhynchus mykiss* (Walbaum). *Journal of Fish Biology* 41:257–263.

McCarthy, I. D., Houlihan, D. F., Carter, C. G., Mouton, K. 1993. Variation in individual food consumption rates of fish and its implications for the study of fish nutrition and physiology. *Proceedings of the Nutrition Society* 52:427–436.

McClure, M. M., McIntyre, P. B., McCune, A. R. 2006. Notes on the natural diet and habitat of eight danionin fishes, including the zebrafish *Danio rerio*. *Journal of Fish Biology* 69:553–570.

McGonnell, I. M., Fowkes, R. C. 2006. Fishing for gene function: Endocrine modelling in the zebrafish. *Journal of Endocrinology* 189:425–439.

McHenry, M. J., Feitl, K. E, Strother, J. A., Van Trump, W. J. 2009. Larval zebrafish rapidly sense the water flow of a predator's strike. *Biology Letters* 5:477–479.

Meinelt, T., Schulz, C., Wirth, M., Kurzinger, H., Steinberg, C. 1999. Dietary fatty acid composition influences the fertilization rate of zebrafish (*Danio rerio* Hamilton–Buchanan). *Journal of Applied Ichthyology* 15:19–23.

Meinelt, T., Schulz, C., Wirth, M., Kurzinger, H., Steinberg, C. 2000. Correlation of diets high in n-6 polyunsaturated fatty acids with high growth rate in zebrafish (*Danio rerio*). *Comparative Medicine* 50(1):43–45.

Milan, D. J., Jones, I. L., Ellinor, P. T., MacRae, C. A. 2006. In vivo recording of adult zebrafish electrocardiogram and assessment of drug-induced QT prolongation. *American Journal of Physiology. Heart and Circulatory Physiology 291*:269–273.

Milewski, W., Duguay, S., Chan, S., Steiner, D. 1998. Conservation of PDX-1 structure, function, and expression in zebrafish. *Endocrinology 139*:1440–1449.

Moens, C. B., Donn, T. M., Wolf-Saxon, E. R., Ma, T. P. 2008. Reverse genetics in zebrafish by TILLING. *Briefings in Functional Genomics and Proteomics 7*:454–459.

Mohammad, R., Hem, A., Shukla, D., Whitaker, B., Arnold, J., Shahamat, M. 2007. Attachment and biofilm formation of *Mycobacterium marinum* on a hydrophobic surface at the air interface. *World Journal of Microbiology and Biotechnology 23*:93–101.

Mölich, A., Waser, W., Heisler, N. 2009. The teleost pseudobranch: A role for preconditioning of ocular blood supply? *Fish Physiology and Biochemistry 35*:273–286.

Moore, F. B., Hosey, M., Bagatto, B. 2006. Cardiovascular system in larval zebrafish responds to developmental hypoxia in a family specific manner. *Frontiers in Zoology 15*:3–4.

Moravec, F., Wolter, J., Korting, W. 1999. Some nematodes and acanthocephalans from exotic ornamental freshwater fishes imported into Germany. *Folia Parasitologica 46*:296–310.

Morris, D. J., Adams, A. 2006. Transmission of freshwater myxozoans during the asexual propagation of invertebrate hosts. *International Journal for Parasitology 36*:371–377.

Moss, J. B., Koustubhan, P., Greenman, M., Parsons, M. J., Walter, I., Moss, L.G. 2009. Regeneration of the pancreas in adult zebrafish. *Diabetes 58*:1844–1851.

Mudumana, S. P., Wan, H., Singh, M., Korzh, V., Gong, Z. 2004. Expression analyses of zebrafish transferrin, ifabp, and elastase B mRNAs as differentiation markers for the three major endodermal organs: Liver, intestine, and exocrine pancreas. *Developmental Dynamics 230*:165–173.

Mukhi, S., Patiño, R. 2007. Effects of prolonged exposure to perchlorate on thyroid and reproductive function in zebrafish. *Toxicological Science 96*:246–254.

Mullins, M. C., Hammerschmidt, M., Haffter, P., Nüsslein-Volhard, C. 1994. Large-scale mutagenesis in the zebrafish: In search of genes controlling development in a vertebrate. *Current Biology* 4:189–202.

Munro, A. D., Scott, P., Lam, T. J., eds. 1990. *Reproductive Seasonality in Teleosts: Environmental Influences.* Boca Raton, FL: CRC Press.

Murtha, J. M., Qi, W., Keller, E. T. 2003. Hematologic and serum biochemical values for zebrafish (*Danio rerio*). *Comparative Medicine* 53:37–41.

Mölich, A., Waser, W., Heisler, N. 2009. The teleost pseudobranch: A role for preconditioning of ocular blood supply? *Fish Physiology and Biochemistry* 35:273–286.

Nagayoshi, S., Hayashi, E., Abe, G., Osato, N., Asakawa, K., Urasaki, A., Horikawa, K. Ikeo, K., Takeda, H., Kawakami, K. 2008. Insertional mutagenesis by the Tol2 transposon-mediated enhancer trap approach generated mutations in two developmental genes: tcf7 and synembryn-like. *Development* 135:159–169.

National Research Council. 1996. *Guide for the Care and Use of Laboratory Animals.* Washington, D.C.: National Academy Press.

Neely, M. N., Pfeifer, J. D., Caparon, M. 2002. Streptococcus-zebrafish model of bacterial pathogenesis. *Infection and Immunity* 70:3904–14.

Nemetz, T. G., Shotts, E. B. 1993. Zoonotic diseases. In *Fish Medicine*, edited by M. K. Stoskopf. Philadelphia: W. B. Saunders.

Nemtsas, P., Wettwer, E., Christ, T., Weidinger, G., Ravens, U. 2009. Adult zebrafish heart as a model for human heart? An electrophysiological study. *Journal of Molecular and Cellular Cardiology.*

Ng, A. N., de Jong Curtain, T. A., Mawdsley, D. J., White, S. J., Shin, J., Appel, B., Dong, P. D., Stainier, D. Y., Heath, J. K. 2005. Formation of the digestive system in zebrafish: III. Intestinal epithelium morphogenesis. *Developmental Biology* 286:114–135.

NIH (National Institutes of Health). 2009. *Final Report to OLAW on Euthanasia of Zebrafish.*

Noga, E. J. 1996. *Fish Disease: Diagnosis and Treatment.* St. Louis: Mosby.

Novak, C. M., Jiang, X., Wang, C., Teske, J. A., Kotz, C. M., Levine, J. A. 2005. Caloric restriction and physical activity in zebrafish (*Danio rerio*). *Neuroscience Letters 383*:99–104.

Nusslein-Volhard, C., Dahm, R. eds. 2002. *Zebrafish, A Practical Approach.* Oxford: Oxford University Press.

Okabe M., Graham A. 2004. The origin of the parathyroid gland. *Proceedings of the National Academy of Sciences of the United States of America 101*:17716–17719.

Olney, C. E., Schauer, P. S. McLean, S., Lu, Y., Simpson, K. L. 1980. *International Study on Artemia. VIII.* Comparison of the chlorinated hydrocarbons and heavy metals in five different strains of newly hatched *Artemia* and a laboratory-reared marine fish. In *The Brine Shrimp Artemia, Vol. 3. Ecology, Culturing, Use in Aquaculture,* edited by G. Persoone, P. Sorgeloos, O. Roels, and E. Jaspers. Wetteren, Belgium: Universa Press.

Olsen, R. E., Sundell, K., Hansen, T., Hemre, G., Myklebust, R., Mayhew, T. M., Ringø, E. 2002. Acute stress alters the intestinal lining of Atlantic salmon, *Salmo salar* L.: An electron microscopical study. *Fish Physiology and Biochemistry 26*:211–221.

Olsen, R. E., Sundell, K., Ringø, E., Mykelbust, R., Hemre, G-I, Hansen, T., Karlsen, Ø. 2008. The acute stress response in fed and food deprived Atlantic cod, *Gadus morhua* L. *Aquaculture 280*:232–241.

Onal, U., Langdon, C. 2000. Characterization of two microparticle types for delivery of food to altricial fish larvae. *Aquaculture Nutrition 6*:159–170.

Onnebo, S. M., Yoong, S. H., Ward, A. C. 2004. Harnessing zebrafish for the study of white blood cell development and its perturbation. *Experimental Hematology 32*:789–796.

Orban, L., Sreenivasan, R., Olsson, P.E. 2009. Long and winding roads: Testis differentiation in zebrafish. *Molecular and Cellular Endocrinology 312*:35–41.

Ortuno, J., Esteban, M. A., Meseguer, J. 2003. The effect of dietary intake of vitamins C and E on the stress response of gilthead seabream (*Sparus aurata* L.). *Fish and Shellfish Immunology 2*:145–156.

Ostland, V. E., Ferguson, H. W., Prescott, J. F., Stevenson, R. W., Barker, I. K. 1990. Bacterial gill disease of salmonids: Relationship between the severity of gill lesions and bacterial recovery. *Diseases of Aquatic Organisms* 9:5–14.

Ostland, V. E., Watral, V., Whipps, C. M., Austin, F. W., St-Hilaire, S., Westerman, M. E., Kent, M. L. 2008. Biochemical, molecular, and virulence characteristics of select *Mycobacterium marinum* isolates in hybrid striped bass *Morone chrysops* x *M. saxatilis* zebrafish *Danio rerio*. *Diseases of Aquatic Organisms* 79:107–118.

Øverli, O., Winberg, S., Pottinger, T. G. 2005. Behavioral and neuroendocrine correlates of selection for stress responsiveness in rainbow trout – A review. *Integrative and Comparative Biology* 45:463–474.

Palumbo, S. A., Bencivengo, M. M., Del Corral, F., Williams, A. C., Buchanan, R. L. 1989. Characterization of the *Aeromonas hydrophila* group isolated from retail foods of animal origin. *Journal of Clinical Microbiology* 27:854–859.

Parichy, D.M. 2006. Evolution of danio pigment pattern development. *Heredity* 97:200–210.

Parmentier, C., Taxi, J., Balment, R., Nicolas, G., Calas, A. 2006. Caudal neurosecretory system of the zebrafish: Ultrastructural organization and immunocytochemical detection of urotensins. *Cell and Tissue Research* 325:111–124.

Patton, E. E., Zon L. I. 2001. The art and design of genetic screens: Zebrafish. *Nature Reviews. Genetics* 2:956–966.

Paw, B. H., Zon, L. I. 2000. Zebrafish: A genetic approach in studying hematopoiesis. *Current Opinion in Hematology* 7:79–84.

Perry, S. F. 1997. The chloride cell: Structure and function in the gills of freshwater fishes. *Annual Review of Physiology* 59:325–347.

Petrie–Hanson, L., Romano, C. L., Mackey, R. B., Khosravi, P., Hohn, C. M., Boyle, C. R. 2007. Evaluation of zebrafish *Danio rerio* as a model for enteric septicemia of catfish (ESC). *Journal of Aquatic Animal Health* 19:151–158.

Petty, B. D., Francis-Floyd, R. 2004. Pet fish care and husbandry. *The Veterinary Clinics of North America. Exotic Animal Practice* 7:397–419.

Phelps, H. A., Runft, D. L., Neely, M. N. 2009. Adult zebrafish model of streptococcal infection. *Current Protocols in Microbiology* Chapter 9: Unit 9D.1.

Pinto, W., Aragão, C., Soares, F., Dinis, M. T., Conceição, L. E. 2007. Growth, stress response and free amino acid levels in Senegalese sole (*Solea senegalensis* Kaup 1858) chronically exposed to exogenous ammonia. *Aquaculture Research 38*:1198–1204.

Pissios, P., Bradley, R., Maratos-Flier, E. 2006. Expanding the scales: The multiple roles of MCH in regulating energy balance and other biological functions. *Endocrine Reviews 27*:606–620.

Popper, A. N., Fay, R. R. 1993. Sound detection and processing by fish: Critical review and major research questions. *Brain, Behavior and Evolution 41*:14–38.

Porazzi, P., Calebiro, D., Benato, F., Tiso, N., Persani, L. 2009. Thyroid gland development and function in the zebrafish model. *Molecular and Cellular Endocrinology 312*:14–23.

Postlethwait, J., ed. 2004. Evolution of the zebrafish genome. In *Fish Development and Genetics: The Zebrafish and Medaka Models*, edited by G. Zhiyuan, and V. Korzh. Singapore: World Scientific Publishing Company.

Postlethwait, J. H. 2006. The zebrafish genome: A review and msx gene case study. *Genome Dynamics 2*:183–197.

Priestley, S. M., Stevenson, A. E., Alexander, L. G. 2006. Growth rate and body condition in relation to group size in black widow tetras (*Gymnocorymbus ternetzi*) and common goldfish (*Carassius auratus*). *The Journal of Nutrition 136*:2078S–2080S.

Pritchard, V. L., Lawrence, J., Butlin, R. K., Krause, J. 2001. Shoal choice in zebrafish, *Danio rerio*: The influence of shoal size and activity. *Animal Behaviour 62*:1085–1088.

Pugach, E. K., Li, P., White, R., Zon, L. 2009. Retro-orbital injection in adult zebrafish. *Journal of Visualized Experiments: JoVE* (34).

Pujic, Z., Malicki, J. 2004. Retinal pattern and the genetic basis of its formation in zebrafish. *Seminars in Cell and Developmental Biology 15*:105–114.

Pullium, J. K., Dillehay, D. L., Webb, S. 1999. High mortality in zebrafish (*Danio rerio*). *Contemporary Topics in Laboratory Animal Science 38*:80–83.

Raftos, D.A. Cooper, E. L. 1990. Diseases of annelids. In *Diseases of Marine Annelids*, edited by O. Kinne. Hamburg: Biologische Anstalt, Helgoland.

Ramsay, J. M., Feist, G. M., Varga, Z. M., Westerfield, M. Kent, M. L., Schreck, C. B. 2009. Whole-body cortisol response of zebrafish to acute net handling stress. *Aquaculture* 297:157–162.

Ramsay, J. M., Watral V., Schreck, C. B., Kent, M. L. 2009. Husbandry stress exacerbates mycobacterial infections in adult zebrafish, *Danio rerio* (Hamilton). *Journal of Fish Diseases* 32:931–941.

Randall, D. J., Tsui, T. K. N. 2002. Ammonia toxicity in fish. *Marine Pollution Bulletin* 45:17–23.

Rawls, J. F., Samuel, B. S., Gordon, J. I. 2004. Gnotobiotic zebrafish reveal evolutionarily conserved responses to the gut microbiota. *Proceedings of the National Academy of Sciences* 101:4596–4601.

Reavill, D. R. 2006. Common diagnostic and clinical techniques for fish. *Veterinary Clinics of North America: Exotic Animal Practice* 9:223–235.

Reichwald, K., Lauber, C., Nanda, I., Kirschner, J., Hartmann, N., Schories, S., Gausmann, U., Taudien, S., Schilhabel, M. B., Szafranski, K., Glöckner, G., Schmid, M., Cellerino, A., Schartl, M., Englert, C., Platzer, M. 2009. High tandem repeat content in the genome of the short-lived annual fish *Nothobranchius furzeri*: A new vertebrate model for aging research. *Genome Biology* 10:R16.

Reimschuessel, R. 2001. A fish model of renal regeneration and development. *ILAR Journal* 42:285–291.

Ren, J. Q., McCarthy, W. R., Zhang, H., Adolph, A. R., Li, L. 2002. Behavioral visual responses of wild-type and hypopigmented zebrafish. *Vision Research* 42:293–9.

Risner, M. L., Lemerise, E., Vukmanic, E. V., Moore, A. 2006. Behavioral spectral sensitivity of the zebrafish (*Danio rerio*). *Vision Research* 46:2625–2635.

Robertson, G. N., Lindsey, B. W., Dumbarton, T. C., Croll, R. P., Smith, F. M. 2008. The contribution of the swimbladder to buoyancy in the adult zebrafish (*Danio rerio*): A morphometric analysis. *Journal of Morphology* 269:666–673.

Robertson, G. N., McGee, C. A., Dumbarton, T. C., Croll, R. P., Smith, F. M. 2007. Development of the swimbladder and its innervation in the zebrafish, *Danio rerio. Journal of Morphology* 268:967.

Robison, B. D., Drew, R. E., Murdoch, G. K., Powell, M., Rodnick, K. J., Settles, M., Stone, D., Churchill, E., Hill, R. A., Papasani, M. R., Lewis, S. S., Hardy, R. W. 2008. Sexual dimorphism in hepatic gene expression and the response to dietary carbohydrate manipulation in the zebrafish (*Danio rerio*). *Comparative Biochemistry and Physiology. D: Genomics & Proteomics* 3:141–154.

Rodríguez, I., Chamorro, R., Novoa, B., Figueras, A. 2009. Beta-glucan administration enhances disease resistance and some innate immune responses in zebrafish (*Danio rerio*). *Fish and Shellfish Immunology* 27:369–373.

Rodríguez, I., Novoa, B., Figueras, A. 2008. Immune response of zebrafish (*Danio rerio*) against a newly isolated bacterial pathogen *Aeromonas hydrophila*. *Fish and Shellfish Immunology* 25:239–249.

Roex, E. W., Keijzers, R., van Gestel, C. A. 2003 Acetylcholinesterase inhibition and increased food consumption rate in the zebrafish, *Danio rerio*, after chronic exposure to parathion. *Aquatic Toxicology* 64:451–460.

Rolfhus, K. R., Sandheinrich, M. B., Wiener, J. G., Bailey, S. W., Thoreson, K. A., Hammerschmidt, C. R. 2008. Analysis of fin clips as a nonlethal method for monitoring mercury in fish. *Environmental Science and Technology* 42:871–877.

Rombough, P. J. 2007. Ontogenetic changes in the toxicity and efficacy of the anaesthetic MS222 (tricaine methanesulfonate) in zebrafish (*Danio rerio*) larvae. *Comparative Biochemistry and Physiology. Part A, Molecular and Integrative Physiology* 148:463–469.

Rombough, P. J. 2002. Gills are needed for ionoregulation before they are needed for O_2 uptake in developing zebrafish, *Danio rerio. The Journal of Experimental Biology* 205:1787–1794.

Rombout, J. H., Lamers, C. H., Helfrich, M. H., Dekker, A., Taverne-Thiele, J. J. 1985. Uptake and transport of intact macromolecules in the intestinal epithelium of carp (*Cyprinus carpio* L.) and the possible immunological implications. *Cell and Tissue Research* 239:519–530.

Rosen, J. N., Sweeney, M. F., Mably, J. D. 2009. Microinjection of zebrafish embryos to analyze gene function. *Journal of Visualized Experiments* 25.

Rothen, D., Curtis, E., Yanong, R. 2002. Tolerance of yolk sac and free-swimming fry of the zebra danio *Brachydanio rerio*, black tetra *Gymnocorymbus ternetzi*, Buenos Aires tetra *Hemigrammus caudovittatus*, and blue gourami *Trichogaster trichopterus* to therapeutic doses of formalin and sodium chloride. *Journal of Aquatic Animal Health 14*:204–208.

Russo, R., Mitchell, H., Yanong, R. 2006. Characterization of *Streptococcus iniae* isolated form ornamental cyprinid fishes and development of challenge models. *Aquaculture 256*:105–110.

Ryba, S. A., Lake, J. L., Serbst, J. R., Libby, A. D., Ayvazian, S. 2008. Assessment of caudal fin clip as a non-lethal technique for predicting muscle tissue mercury concentrations in largemouth bass. *Environmental Chemistry 5*:200–203.

Saha, N.C., Bhunia, F., Kaviraj, K. 1999. Toxicity of phenol to fish and aquatic ecosystems. *Bulletin of Environmental Contamination and Toxicology 63*:195–202.

Sakamoto, T., McCormick, S. D. 2006. Prolactin and growth hormone in fish osmoregulation. *General and Comparative Endocrinology 147*:24–30.

Sanders, G. E., Batts, W. N., Winton, J. R. 2003. Susceptibility of zebrafish (*Danio rerio*) to a model pathogen, spring viremia of carp virus. *Comparative Medicine 53*:514–521.

Sanders, L. H., Whitlock, K. E. 2003. Phenotype of the zebrafish masterblind (mbl) mutant is dependent on genetic background. *Developmental Dynamics 227*:291–300.

Sawant, M. S.,Zhang, S., Li, L. 2001. Effect of salinity on development of zebrafish, *Brachydanio rerio*. *Current Science 81*:1347–1350.

Schachte, J. H. 1983. Bacterial gill disease. In *Guide to Integrated Fish Health Management in the Great Lakes Basin*, edited by F. P. Meyer, J. W. Warren, and T. G. Carey. Ann Arbor: Great Lakes Fishery Commission, Special Publication.

Schaefer, J., Ryan, A. 2006. Developmental plasticity in the thermal tolerance of zebrafish *Danio rerio*. *Journal of Fish Biology 69*(4):722–734.

Schlueter, P. J., Royer, T., Farah, M. H., Laser, B., Chan, S. J., Steiner, D. F., Duan, C. 2006. Gene duplication and functional divergence of the zebrafish insulin-like growth factor 1 receptors. *FASEB Journal 20*:1230–32.

Schreck, C. B. 2009. Stress and fish reproduction: The roles of allostasis and hormesis. *General and Comparative Endocrinology 165*:549–556.

Schultz, I. R., Reeda, S., Pratta, A., Skillmana, A. D. 2007. Quantitative oral dosing of water soluble and lipophilic contaminants in the Japanese medaka (*Oryzias latipes*). *Comparative Biochemistry and Physiology Part C: Toxicology and Pharmacology 145*:86–95.

Scott, A. L., Rogers, W.A. 2006. Histological effects of prolonged sublethal hypoxia on channel catfish *Ictalurus punctatus* (Rafinesque). *Journal of Fish Diseases 3*:305–316.

Scott, A. P., Ellis T. 2007. Measurement of fish steroids in water – A review. *General and Comparative Endocrinology 153*:392–400.

Selman, K.,Wallace, R. A., Sarka, A., Qi, X. P. 1993. Stages of oocyte development in the zebrafish *Brachydanio rerio*. *Journal of Morphology 218*:203–224.

Sessa, A. K., White, R., Houvras, Y., Burke, C., Pugach, E., Baker, B., Gilbert, R., Look, A. T., Zon, L. I. 2008. The effect of a depth gradient on the mating behavior, oviposition site preference, and embryo production in the zebrafish, *Danio rerio*. *Zebrafish 5*:335–339.

Shen, C. H., Steiner, L. A. 2004. Genome structure and thymic expression of an endogenous retrovirus in zebrafish. *Journal of Virology 78*:899–911.

Shih, T. H., Horng, J. L., Hwang, P. P., Lin, L. Y. 2008. Ammonia excretion by the skin of zebrafish (*Danio rerio*) larvae. *American Journal of Physiology: Cell Physiology 295*:1625–1632.

Shoemaker, C., Klesius, P. 1997. Streptococcal disease problems and control: A review. In *Tilapia Aquaculture*, edited by K. Fitzsimmons. Ithaca, NY: Northeast Regional Agricultural Engineering Service 106.

Siccardi, A. J., Garris, H. W., Jones, W. T., Moseley, D. B., D'Abramo, L. R., Watts, S. A. 2009. Growth and survival of zebrafish (*Danio rerio*) fed different commercial and laboratory diets. *Zebrafish 6*:275–280.

Silverman, J., Suckow, M. A., Murphy, S. 2006. *The IACUC Handbook*. 2nd ed. Boca Raton, FL: Taylor & Francis.

Singh, S. K. 2007. Endogenous retroviruses: Suspects in the disease world. *Future Microbiology 2*:269–275.

Sitjà-Bobadilla, A. 2008. Fish immune response to Myxozoan parasites. *Parasite 15*:420–425.

Song, Y., Cone, R. D. 2007. Creation of a genetic model of obesity in a teleost. *The FASEB Journal 21*:2042–2049.

Soules, K. A., Link, B. A. 2005. Morphogenesis of the anterior segment in the zebrafish eye. *BMC Developmental Biology 5*:12.

Spence, R., Ashton, R., Smith, C. 2007. Adaptive oviposition choice in the zebrafish, *Danio rerio. Behaviour 144*:953–966.

Spence, R., Ashton, R. L., Smith, C. 2007. Adaptive oviposition decisions are mediated by spawning site quality in the zebrafish, *Danio rerio. Behaviour 144*:953–966.

Spence, R., Fatema, M. K., Ellis, S., Ahmed, Z. F., Smith, C. 2007. Diet, growth and recruitment of wild zebrafish in Bangladesh. *Journal of Fish Biology 71*:304–309.

Spence, R., Fatema, M. K., Reichard, M., Huq, K. A., Wahab, M. A., Ahmed, Z. F., Smith, C. 2006. The distribution and habitat preferences of the zebrafish in Bangladesh. *Journal of Fish Biology 69*:1435–1448.

Spence, R., Fatema, M. K., Ellis, S., Ahmed, Z. F., Smith, C. 2007. The diet, growth and recruitment of wild zebrafish (*Danio rerio*) in Bangladesh. *Journal of Fish Biology 71*:304–309.

Spence, R., Gerlach, G., Lawrence, C., Smith, C. 2008. The behaviour and ecology of the zebrafish, *Danio rerio. Biological Reviews 83*:13–34.

Spence, R., Smith, C. 2005. Male territoriality mediates density and sex ratio effects on oviposition in the zebrafish, *Danio rerio. Animal Behaviour 69*:1317–1323.

Spence, R., Smith, C. 2006. Mating preference of female zebrafish, *Danio rerio*, in relation to male dominance. *Behavioral Ecology 17*:779–783.

Spence, R., Smith, C. 2007. The role of early learning in determining shoaling preferences based on visual cues in the zebrafish, *Danio rerio. Ethology 113*:62–67.

Spencer, P., Pollock, R., Dubé, M. 2008. Effects of un-ionized ammonia on histological, endocrine, and whole organism endpoints in slimy sculpin (*Cottus cognatus*). *Aquatic Toxicology 90*:300–309.

Stock, D. W. 2007. Zebrafish dentition in comparative context. *Journal of Experimental Zoology Part B: Molecular and Developmental Evolution 308*:523–549.

Stockhammer, O. W., Zakrzewska, A., Hegedûs, Z., Spaink, H. P., Meijer, A. H. 2009. Transcriptome profiling and functional analyses of the zebrafish embryonic innate immune response to Salmonella infection. *Journal of Immunology 182*:5641–5653.

Stohler, R. A., Curtis, J., Minchella, D. J. 2004. A comparison of microsatellite polymorphism and heterozygosity among field and laboratory populations of *Schistosoma mansoni. International Journal for Parasitology 34*:595–601.

Stoskopf, M. K. 1993. *Fish Medicine*. Philadelphia: W. B. Saunders Co.

Sullivan, C., Kim, C. H. 2008. Zebrafish as a model for infectious disease and immune function. *Fish and Shellfish Immunology 25*:341–350.

Summerfelt, R. C., Smith, L. S. 1990. Anaesthesia, surgery and related techniques. In *Methods for Fish Biology*, edited by C. B. Schreck and P. B. Moyle. Bethesda, MD: American Fisheries Society.

Swann, L. 1997. A fish farmer's guide to understanding water quality in aquaculture. Illinios–Indiana Sea Grant Program, Aquaculture Extension.

Sørum, U., Toften, H., Damsgård, B. *Feed Intake, Health and Behavior as Fish Welfare Indicators*. Available from http://www.mun.ca/virt/*aquanet*/English/research/fish/sorum.pdf.

Talwar, P. K., Jingran, A. G. 1991. *Inland Fishes of India and Adjacent Countries*. Rotterdam: A. A. Balkema.

Tan, Y., Sun, D., Huang, W., Cheng, S. H. 2008. Mechanical modeling of biological cells in microinjection. *IEEE Trans Nanobioscience 7*:257–266.

Tanimoto, M., Ota, Y., Horikawa, K., Oda, Y. 2009. Auditory input to CNS is acquired coincidentally with development of inner ear after formation of functional afferent pathway in zebrafish. *The Journal of Neuroscience 29*:2762–2767.

Thomas, P., Rahman, M. S., Khan, I. A., Kummer, J. A. 2007. Widespread endocrine disruption and reproductive impairment in an estuarine fish population exposed to seasonal hypoxia. *Proceedings of the Royal Society B 274*:2693–2701.

Timmons, M. B., Ebeling, J. M., Wheaton, J. M., Summerelt, S. T., Vinci, B. J., eds. 2002. *Recirculating Aquaculture Systems*. 2nd ed. Ithaca, NY: Cayuga Aqua Ventures.

Toyama, R., Chen, X., Jhawar, N., Aamar, E., Epstein, J., Reany, N., Alon, S., Gothilf, Y., Klein, D. C., Dawid, I. B. 2009. Transcriptome analysis of the zebrafish pineal gland. *Developmental Dynamics* 238:1813–1826.

Treasurer, J. W. 2006. Measurement of regurgitation in feeding studies of predatory fishes. *Journal of Fish Biology* 33:267–271.

Trenzado, C. E., Morales, A. E., Higuera, M. 2008. Physiological changes in rainbow trout held under crowded conditions and fed diets with different levels of vitamins E and C and highly unsaturated fatty acids (HUFA). *Aquaculture* 277:293–302.

Trevarrow, B., Robison, B. 2004. Genetic backgrounds, standard lines, and husbandry of zebrafish. *Methods in Cell Biology* 77:599–616.

Tsai, S. B., Tucci, V., Uchiyama, J., Fabian, N. J., Lin, M. C., Bayliss, P. E., Neuberg, D. S., Zhdanova, I. V., Kishi, S. 2007. Differential effects of genotoxic stress on both concurrent body growth and gradual senescence in the adult zebrafish. *Aging Cell* 6:209–224.

Tseng, D. Y., Chou, M. Y., Tseng, Y. C., Hsiao, C. D., Huang, C. J., Kaneko, T., Hwang, P. P. 2009. Effects of stanniocalcin 1 on calcium uptake in zebrafish (*Danio rerio*) embryo. *American Journal of Physiology. Regulatory, Integrative and Comparative Physiology* 296:549–557.

Turner, J. W. Jr., Nemeth, R., Rogers, C. 2003. Measurement of fecal glucocorticoids in parrotfishes to assess stress. *General and Comparative Endocrinology* 133:341–352.

van der Meer, D. L., van den Thillart, G. E., Witte, F., de Bakker, M. A., Besser, J., Richardson, M. K., Spaink, H. P., Leito, J. T., Bagowski, C. P. 2005. Gene expression profiling of the long-term adaptive response to hypoxia in the gills of adult zebrafish. *American Journal of Physiology. Regulatory, Integrative and Comparative Physiology* 289:R1512–1519.

Vandenhurk, R., Lambert, J. G. D. 1983. Ovarian-steroid glucuronides function as sex-pheromones for male zebrafish, *Brachydanio rerio*. *Canadian Journal of Zoology – Revue Canadienne de Zoologie* 61:2381–2387.

Volpato, G. L. 2009. Challenges in assessing fish welfare. *ILAR Journal* 50:329–337.

Volpato, G. L., Barreto, R. E. 2001. Environmental blue light prevents stress in the fish Nile tilapia. *Brazilian Journal of Medical and Biological Research 34*:1041–1045.

Wagner, G. N., Stevens, E. D., Harvey-Clark, C. J. 1999. Wound healing in rainbow trout (*Onchorhyncus mykiss*) following surgical site preparation with a povidone-iodine antiseptic. *Journal of Aquatic Animal Health 11*:373–382.

Walker, C., Streisinger, G. 1983. Induction of mutations by gamma-rays in pre-gonial germ-cells of zebrafish embryos. *Genetics 103*:125–136.

Wallace, K. N., Akhter, S., Smith, E. M., Lorent, K., Pack, M. 2005. Intestinal growth and differentiation in zebrafish. *Mechanisms of Development 122*:157–173.

Wang, D., Jao, L. E., Zheng, N., Dolan, K., Ivey, J., Zonies, S., Wu, X., Wu, K., Yang, H., Meng, Q., Zhu, Z., Zhang, B., Lin, S., Burgess, S. M. 2007. Efficient genome-wide mutagenesis of zebrafish genes by retroviral insertions. *Proceedings of the National Academy of Sciences of the United States of America 104*:12428–12433.

Wang, S., Yuen, S. S., Randall, D. J., Hung, C. Y., Tsui, T. K., Poon, W. L., Lai, J. C., Zhang, Y., Lin, H. 2008. Hypoxia inhibits fish spawning via LH-dependent final oocyte maturation. *Comparative Biochemistry and Physiology C: Comparative Pharmacology and Toxicology 148*:363–369.

Wang, Z., Zhang, S., Tong, Z., Li, L., Wang, G. 2009. Maternal transfer and protective role of the alternative complement components in zebrafish *Danio rerio. PLoS One 4*:e4498.

Waser, W., Heisler, N. 2005. Oxygen delivery to the fish eye: Root effect as crucial factor for elevated retinal PO2. *The Journal of Experimental Biology 208*(Pt 21):4035–4047.

Weber, E. S., Innis, C. 2007. Piscine patients: Basic diagnostics. *Compendium 29*:276–277.

Wedemeyer, G. A. 1996. *Physiology of Fish in Intensive Culture Systems*. New York: Chapman & Hall.

Weinstein, M. R., Litt, M., Kertesz, D. A., Wyper, P., Rose, D., Coulter, M., McGeer, A., Facklam, R., Ostach, C., Willey, B. M., Borczyk, A., Low, D. E. 1997. Invasive infections due to a fish pathogen, *Streptococcus iniae. S. iniae* study group. *The New England Journal of Medicine 337*:589–594.

Welker, T., Lim, C., Aksoy, M., Klesius, P. 2007. Effect of buffered and unbuffered tricaine methanesulfonate (MS–222) at different concentrations on the stress responses of channel catfish (*Ictalurus Punctatus Rafinesque*). *Journal of Applied Aquaculture* 19:1–18.

Wendl, T., Lun, K., Mione, M., Favor, J., Brand, M., Wilson, S. W., Rohr, K. B. 2002. Pax2.1 is required for the development of thyroid follicles in zebrafish. *Development* 129:3751–3760.

Westerfield, M. 2007. *The Zebrafish Book: A Guide for the Laboratory Use of Zebrafish.* 4th ed. Eugene: University of Oregon Press.

Wheaton, F. 2002. Recirculating aquaculture systems: An overview of waste management. Paper presented at Proceedings of the Fourth International Conference on Recirculating Aquaculture.

Wheaton, F. W., Hochheimer, J. N., Kaiser, G. E., Malone, R. F., Krones, M. J., Libey, G. S., Easter, C. 1994. Nitrification filter design methods. In *Aquaculture Water Reuse Systems: Engineering Design and Management,* edited by M. B. Timmons and T. M. Losordo. Amsterdam: Elsevier Science.

Whipps, C. M., Dougan, S. T., Kent, M. L. 2007. *Mycobacterium haemophilum* infections of zebrafish (*Danio rerio*) in research facilities. *FEMS Microbiology Letters* 270:21–26.

Whipps, C. M., Kent, M. L. 2006. Polymerase chain reaction detection of *Pseudoloma neurophilia,* a common microsporidian of zebrafish (*Danio rerio*) reared in research laboratories. *Journal American Association Laboratory Animal Science* 45:36–39.

Whipps, C. M., Matthews, J. L., Kent, M. L. 2008. Distribution and genetic characterization of *Mycobacterium chelonae* in laboratory zebrafish *Danio rerio. Diseases of Aquatic Organisms* 82:45–54.

Whitfield, T. T., Riley, B. B., Chiang, M. Y., Phillips, B. 2002. Development of the zebrafish inner ear. *Developmental Dynamics* 223:427–458.

Wilkie, M. P. 2002. Ammonia excretion and urea handling by fish gills: Present understanding and future research challenges. *The Journal of Experimental Zoology* 293:284–301.

Willett, C. E., Cortes, A., Zuasti, A., Zapata, A. G. 1999. Early hematopoiesis and developing lymphoid organs in the zebrafish. *Developmental Dynamics* 214:323–336.

Wilson, C. A. 2008. Porcine endogenous retroviruses and xenotransplantation. *Cellular and Molecular Life Sciences* 65:3399–3412.

Wilson, J. M., Randall, D. J., Donowitz, M., Vogl, A. W., Ip, A. K. 2000. Immunolocalization of ion-transport proteins to branchial epithelium mitochondria-rich cells in the mudskipper (*Periophthalmodon schlosseri*). *The Journal of Experimental Biology 203*:2297–2310.

Wilson, J. M., Bunte, R. M., and Carty, A. J. 2009. Evaluation of rapid cooling and tricaine methanesulfonate (MS–222) as a methods of eunthanasia in zebrafish (*Danio rerio*). *Journal American Association Laboratory Animal Science 48*:785–789.

Wingert, R. A., Davidson A. J. 2008. The zebrafish pronephros: A model to study nephron segmentation. *Kidney International 73*:1120–1127.

Wingert, R. A., Selleck, R., Yu, J., Song, H. D., Chen, Z., Song, A., Zhou, Y., Thisse, B., Thisse, C., McMahon, A. P., Davidson, A. J. 2007. The cdx genes and retinoic acid control the positioning and segmentation of the zebrafish pronephros. *PLoS Genetics 10*:1922–1938.

Wooster, G. A., Hsu, H. M., and Bowser, P. R. 1993. Nonlethal surgical procedures for obtaining tissue samples for fish health inspections. *Journal of Aquatic Animal Health 5*:157–164.

Wright, D., Rimmer, L. B., Pritchard, V. L., Krause, J., Butlin, R. K. 2003. Inter- and intra-population variation in shoaling and boldness in the zebrafish (*Danio rerio*). *Naturwissenschaften 90*:374–377.

Wurts, W. A. 1993. Understanding water hardness. *World Aquaculture 24*:1–18.

Wurts, W. A. 2002. Alkalinity and hardness in production ponds. *World Aquaculture 33*:16–17.

Xu, H., Yang, J., Wang, Y., Jiang, Q., Chen, H., Song, H. 2008. Exposure to 17alpha-ethynylestradiol impairs reproductive functions of both male and female zebrafish (*Danio rerio*). *Aquatic Toxicology 88*:1–8.

Xu, Q. 1999. Microinjection into zebrafish embryos. In *Molecular Embryology Methods and Protocols*, edited by M. Guille. Totowa, NJ: Humana Press.

Yanong, R. 2003. Fish Health Management Considerations in Recirculating Aquaculture Systems – Part 1: Introduction and General Principles. In *Circular FA–120*: UF IFAS Cooperative Extension Service.

Yanong, R. 2003. Fish Health Management Considerations in Recirculating Aquaculture Systems – Part 2: Pathogens. In *Circular FA–121*: UF IFAS Cooperative Extension Service.

Yee, N. S., Lorenta, K., Packa, M. 2005. Exocrine pancreas development in zebrafish. *Developmental Biology 284*:84–101.

Yokogawa, T., Marin, W., Faraco, J., Pezeron, G., Appelbaum, L., Zhang, J., Rosa, F., Mourrain, P., Mignot, E. 2007. Characterization of sleep in zebrafish and insomnia in hypocretin receptor mutants. *PloS Biology 5*:2379–2397.

Yuan, S., Sun, Z. 2009. Microinjection of mRNA and morpholino antisense oligonucleotides in zebrafish embryos. *Journal of Visualized Experiments 25*.

appendix 1

sample *Artemia* decapsulation SOP

1. Hydrate cysts in water (1 g/30 mL water) for 60–90 minutes with moderate aeration.

2. Filter (70–90 μm mesh) cysts and place in container with pre-chilled buffer solution (0.3 mL 40% NaOH + 4.7 mL seawater (35 ppt – ~41 g aquarium salt mix/L reverse osmosis water per g of cysts).*

3. Add 10 mL of liquid bleach per g of cysts. Stir and aerate cysts continuously until color change reaches a bright orange. Color will progress from dark brown to gray, to white and finally orange (though this depends on where cysts come from: e.g., argent platinum cysts turn a dark maroonish color). Process usually takes 5–10 minutes. A good rule of thumb is to go 1–3 minutes beyond the final color change.

4. Filter cysts again and rinse with water.

5. Wash cysts in 0.6% acetic acid for 1 minute to neutralize bleach.

6. Decapsulated cysts may be hatched immediately or stored at 4°C for up to 7 days before hatching. They can also be stored for longer periods in hypersaline (~15–20%) solution at 4°C.**

7. To hatch cysts, place in 5–10 ppt water and aerate overnight. Add calcium carbonate or sodium bicarbonate (0.2 g/L) to

bring to pH > 8. Recommended density of cysts in hatching cone is 2.5–5 g/L.

8. To feed, drain entire contents of tower into filter, rinse gently, and then add to squirt bottles with fish water.

The NaOH solution and 35 ppt water should be kept in separate containers and stored at 4°C.

**2–3 hours after first placing cysts in hypersaline, pour off old solution and replace with fresh hypersaline solution to ensure that cysts remain dehydrated during storage.*

To make acid

10% solution

450 mL DI water

50 mL glacial acetic acid

0.6% solution

940 mL DI water

60 mL 10% acetic acid

appendix 2

Sample Zebrafish Facility Checksheet

	Mon	Tues	Wed	Thurs	Fri	Sat	Sun
System 1							
Temperature (C°)							
Conductivity (uS)							
pH							
NH_4/NH_3 (ppm)							
NO_2 (ppm)							
NO_3 (ppm)							
H_2O meter reading							
Comments							
System 2							
Temperature (C°)							
Conductivity (uS)							
pH							
NH_4/NH_3 (ppm)							
NO_2 (ppm)							
NO_3 (ppm)							
H_2O meter reading							
Comments							
System 3							
Temperature (C°)							
Conductivity (uS)							
pH							
NH_4/NH_3 (ppm)							
NO_2 (ppm)							
NO_3 (ppm)							
H_2O meter reading							
Comments							

	Mon	**Tues**	**Wed**	**Thurs**	**Fri**	**Sat**	**Sun**
System 4							
Temperature (C°)							
Conductivity (uS)							
pH							
NH$_4$/NH$_3$ (ppm)							
NO$_2$ (ppm)							
NO$_3$ (ppm)							
H$_2$O meter reading							
Comments							

Index